VISIBLE BONES

VISIBLE BONES

*Journeys Across Time in the
Columbia River Country*

JACK NISBET

SASQUATCH BOOKS
SEATTLE

AUTHOR'S NOTE ON GEOGRAPHY AND LANGUAGE:
In this book the Columbia Plateau refers to the part of the river's drainage that lies between the Rocky Mountains and the Cascade Range. The Columbia Basin denotes the arid central portion of the Plateau. The Snake River is part of the larger Columbia drainage.

Ethnologists divide the native peoples of the Columbia drainage into three cultural groups. Plateau tribes inhabited most of the interior and spoke Kootenai, Interior Salish, or Sahaptin languages. Great Basin peoples, all Shoshoean speakers, were concentrated in the Snake River country. The Coastal cultures along the lower Columbia spoke Chinookan and Coast Salish languages.

Printed in the United States of America
Published by Sasquatch Books
Distributed by PGW/ Perseus
12 11 10 09 08   8 7 6 5 4 3 2

Cover design: Bob Suh
Interior design and composition: Stewart A. Williams
Cover map: David Thompson. "Map of North America from 84° West," 1804. Map courtesy of
        The National Archives, Kew, United Kingdom.
Cover illustration: John Kirk. Ornithology of the United States of North America. Plate 1. 1839,
        "California Vulture." Illustration courtesy of the Academy of Natural Sciences, Ewell Sale
        Stewart Library, Philadelphia, Pennsylvania.
All interior illustrations by Marjorie C. Leggitt except for:
Page 41: "Head of a Vulture" from the Journal of Meriwether Lewis, February 17, 1806,
        Codex J. p. 80.
Production editor: Heidi A. Schuessler
Copy editor: Don Graydon

Library of Congress Cataloging in Publication Data

Nisbet, Jack, 1949—
        Visible bones : journeys across time in the Columbia River country / Jack Nisbet.
    p. cm.
ISBN 1-57061-376-1 (hardcover)
ISBN 10: 1-57061-524-1 / ISBN 13: 978-1-57061-524-5 (paperback)
Columbia River Valley—Description and travel. 2. Columbia River Valley—History.. 3. Columbia
River Valley—Antiquities. 4. Natural History—Columbia River Valley. 5. Fossils—Columbia River
Valley. I. Title.

F853.N57 2003
979.7—dc21

2003045609

Sasquatch Books
119 South Main Street, Suite 400, Seattle, WA 98104
(206) 467-4300
www.sasquatchbooks.com
custserv@sasquatchbooks.com

Contents

THE COLUMBIA RIVER COUNTRY

Introduction

THE MOMENT THE REAR WHEEL BROKE through the crust, I knew I was really stuck. Muttering curses, I switched off the engine and surveyed the situation. I had backed up too near a grove of birch trees that surrounded a seeping spring, and one wheel was buried to the hubcap. The slice of blue clay that had enveloped my tire emitted a whiff of indigo perfume, and when I bent to look I caught a glimpse of something long and smooth embedded in the mud. At first I took it for the leg bone of an animal, and struggled to pry it free—I've always loved to pick up bones and try to figure out what animal they belonged to—but the fragment proved to be nothing more than a plank from a farmer's old spring box, warped to a pleasing arc by the preserving goo. Rubbing my fingers over the raised grain of the board, I figured that it must have been only a century or so since the birch spring homesteader had set the board in place.

Now the rut that had dragged down my car was offering a small relic from those days for my perusal. The spring where I was mired lay on an open bench, and I walked out to its edge and looked down on the Columbia River. From the site of my misfortune, the whole country was laid out for me to see—open hillsides of ponderosa pine, new outcrops of colored dolomite, draws filled with darker Douglas firs and yellow-green tamarack. Time marched backward from the homestead spring, to the fur traders who had floated past, riding the initial wave of European contact, to the tribal memories buried beneath the waters. The round peaks to the north still capped with snow in early June hinted at the glaciers that had carved the bench where I stood.

The word *relic* conjures up a host of connotations, from human remains to a historic souvenir. It can denote a custom from the past, the remnants of an ancient language, or a fragment of a whole. It can represent the last of a dying species, or an indefatigable survivor. During the years I have lived in the Columbia country,

I have come to see its vast natural and human archives as a reliquary of its former lives, a reservoir of clues that connect this moment to the distant past, this place to territories far away.

That is where this book begins, with the discovery of such relics. Some of them fit comfortably in a pocket; some are far too large for transport. Certain ones have mesmerized generations of chroniclers and invited intense scientific scrutiny, while others have barely been noticed. Many take the tangible form of rock or bone; others are as ephemeral as the faint whiff of a bloom in spring or the laugh of an auntie poking fun. Some have faded to extinction; others can be found by any kid with a penchant for muddy feet. But whatever form they take, each evokes a facet of the region's past, reminding us that this place has not always been as we see it now. Their voices call across time, carrying snatches of the big river's long and larger song.

viii

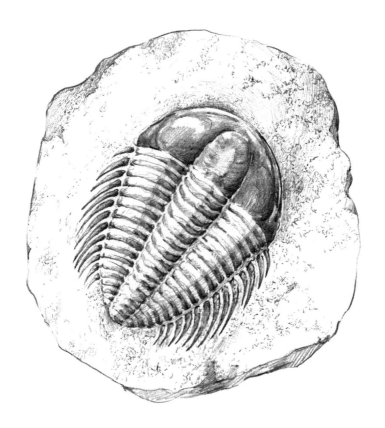

Little Stone House

Upper Cambrian trilobite *(Labiostria westropi)* from Tanglefoot Creek

Tanglefoot ~

"LOOK FOR COOKIES," Rolf had said, as he directed my attention to a tiny squiggle on a map of southeastern British Columbia. "Little round treats in the streambed." He made it sound so easy.

By the time I reached Tanglefoot Creek on the west slope of the Rockies, I was beginning to wish I had waited until after spring runoff. Rising temperatures had loosened the snowpack from the nearby peaks, and the world seemed to be collapsing all around. A grinding porridge of mud and gravel sluiced bushes from steep rock faces. The creek, milky green in color, snapped like a racer snake, carrying chunks of my trail headlong toward the sea.

When I paused to check my progress, heavy drops of rain spanked down on my scrawled directions. I turned up a side rivulet that careened through a tight canyon, picking my way across a fresh mudslide. Clumps of saxifrage flowers, white stars touched with maroon and lemon spots, surfed atop thin plates of brown shale. Outcrops of the same shale shot steeply upward on either side of the creek; this was country that had been bent and twisted on a grand scale. Each step forward in space moved me backward in time.

The narrowing canyon finally forced me into the stream. I swapped boots for water sandals and plunged into the torrent. The water was so cold that I had to hop onto a boulder every few minutes to let the sting go out of my feet. Grabbing at gooseberry bushes to steady myself, I cast my eyes left and right to match the pace of the torrid runoff, searching for the remains of a creature long extinct. The first round stone I picked up turned out to be completely smooth. So did the next several dozen. A thrush's song ascended leisurely over the creek's icy roar, and a succession of hard showers rode in one upon the next. The bird sang many times before an emerging ray of sunlight caught the raised edge of a biscuit-shaped rock on the nose of a gravel bar. I bent down and wrapped my hand around it, feeling for ridges. Even before I lifted it free of the creek, my fingers told me I had found a trilobite.

I waded over to the bank and sat down to admire my prize. It proved to be a worn, warped specimen not much bigger than an Oreo. The three lobes that had once defined a living trilobite were squashed almost flat. The ribbed segments of its thorax showed only as black shadows on the dark green rock, and the code of spiny detail had been reduced to faint cracks. Yet as I squeezed the patterned stone, my body flooded with warmth. Under the spell of the thrush's song, the ancient relic began to spin a tune all its own.

Trilobite Nation ~

IT WAS A TALE that began long ago, back in a time when life existed only in the sea. Beneath the surface of a placid ocean that lapped at the edge of our ancestral continent, the trilobite riffled through the mud. Quill-like spines curved backward along the sides of its squat body. Multiple pairs of jointed legs propelled it forward. As it moved, articulated hinges along its back flexed and rippled like slats on a rolltop desk, and feathered gills along its upper legs combed oxygen from the water. Supple antennae twisted above its head, sensing the surroundings through fine lateral hairs. A pair of prismatic eyes bulged from its rounded head, keen enough to catch movements through the murky depths. A host of images would have flashed across those ancient eyeballs, for the trilobite's home teemed with life. Spiny sponges and pedestaled brachiopods bloomed across the ocean floor, while exotic jellyfish floated in the water column, and segmented worms writhed through the mud. A medley of arthropods, with their jointed limbs and tough outer shells, scrabbled about. Other trilobites, more than a dozen species of them, fanned out across the seafloor habitats like wood warblers through a hardwood forest.

My little trilobite would have begun life as a pin-sized larva drifting in this sea. The tiny creature soon developed a hard calcite carapace that shielded its body. As the animal grew, that protective shell became tighter and tighter, until wrinkled sutures atop its head softened, then cracked open like a locust's shell. Plates around the eyes and cheeks broke free, and the trilobite began to lever its way through the opening. Once released, it was as vulnerable as a soft-shelled crab until a new suit of armor came of age. Over the course of its life, the trilobite discarded many more shields, each slightly larger than the last. When death claimed the animal, its carcass joined those molted shells on the ocean floor. Far beneath the reach of waves and wind, bacteria converged to consume its soft body parts. A gentle shower of silt soon covered the empty shell with a blanket of fine mud.

Time passed. Rivers continued to sluice sediments into the sea. Inch upon inch of primordial goo sifted down atop the trilobite's shield, and myriad ones around it, until they were buried thousands of feet deep. The accumulating weight of all that sediment flattened the trilobite's skeleton and pressed the moisture from the layered silt. As mud was transformed into rock, a peculiar chemical reaction took place between the calcium in the trilobite's exoskeleton and minerals in the surrounding mudstone. Crystals of calcite sprouted around the carapace, forming a rounded nodule with the trilobite's shape perfectly replicated on its surface, as if embossed with the state seal of some ancient arthropodean republic.

Meanwhile, far above its crystalline sarcophagus, my trilobite's kin still crawled, and the remains of countless more generations collected on the seafloor. As the oceans grew colder near the close of the Cambrian period, half a billion years ago, many long-established varieties faded into extinction. New families came into prominence, along with familiar forms of starfish, cuttlefish, bivalved clams, and corals. Jawless fish gave way to sharks, and primitive vegetation appeared on shore. Continents began drifting together to form Pangaea, sea levels rose and fell, climates warmed and cooled and warmed again. Insects took to the air, and amphibians established themselves on solid ground. In the sea, a different and less diverse suite of trilobites scuttled next to horseshoe crabs.

Then, around 250 million years ago, at the end of the Permian age, trilobites disappeared from the seas of our world. They had been part of the saltwater scene for over 350 million years, and then they were gone. Some of their arthropod relatives survived, and their distant cousin the horseshoe crab is with us still, but the trilobite tribe left no direct descendants. The entire evidence of their existence lay locked in vast stone cemeteries thousands of feet beneath the sea.

Tens of millions of years passed before Pangaea began to split apart, and the mechanics of continental drift triggered a series of tectonic collisions off the western coast of North America. Secure within its crypt, my trilobite was slowly nudged ashore. Millimeter by millimeter, it traveled hundreds of miles eastward and thousands of feet upward as the old seabed became a new mountain range. Many more years of grinding ice and rushing water exposed the seam of fossil-laden shale. At the twilight of the last great glacial epoch, the Kootenay and Columbia Rivers settled into the courses we see them run today, carrying the Tanglefoot's flow from the west slope of the Rockies to the Pacific. Birds migrated north and south along the ridgetops, and herding mammals wore pathways back and forth across the Continental Divide. In time, people followed.

The rising waters of formal science did not touch the eroding shale up the Tanglefoot until the late 1950s, when a graduate student stumbled upon some fossils while doing fieldwork in the area. He and subsequent geologists described a lagerstätten—a trove of beautifully preserved specimens spilling out in an abundance that echoed that of the primordial sea. Paleontologists working at the site have since collected thousands of trilobites belonging to over a dozen different species, including two completely new to science.

Stone House~

THE GRADUATE STUDENT, it seems, was not the first visitor to pick up a Tanglefoot trilobite. Several years ago, a retired school-teacher from southwestern British Columbia donated a collection of artifacts to a local confederation of Coast Salish tribes. The items, gathered along the lower Fraser River, included projectile points, scrapers, and knives; the tribal archaeologist noted that several of the pieces were of a sort associated with traditional burial sites. Present in the array was a biscuit-shaped stone that contained some kind of fossil. Rolf Ludvigsen, a paleontologist who directs a research institute in western B.C., was called in to have a look.

Ludvigsen instantly recognized a trilobite of a very unusual type. Furthermore, he knew the species, with its distinctive method of preservation, was found in only one place—Tanglefoot Creek, clear on the opposite side of the province, fully three hundred miles east. But the Tanglefoot belongs to the Columbia drainage, and there is no natural force that could explain how the fossil crossed to the Fraser River system. It could only have been transported across the watersheds by human hands. Ludvigsen speculated that the trilobite might have been picked up by a native traveler who either carried it on a long journey or introduced it into a trading network that eventually led to the lower Fraser. As a student of trilobite lore as well as morphology, he knew that such an occurrence was not without precedent.

Trilobite fossils are found on every continent, and the annals of archaeology hold evidence that these stone images have been catching the eyes of humans since Paleolithic times. An aboriginal tool uncovered in Australia had been chipped from a piece of chert containing a complete trilobite that retained enough distinguishing features to be identified as a new species.

At a rock shelter in central France now known as La Grotte du

6

Trilobite, archaeologists excavating a layer of debris occupied by humans around fifteen thousand years ago unearthed an oblong facsimile of a beetle, carved from lignite coal. Near the beetle lay a worn trilobite. Both artifacts matched recognizable species of the Arthropod order, and both were perforated by carefully placed holes, presumed to have carried a string so that the ornaments could hang in necklace fashion. There is no way to know what these objects meant to their ancient crafters, but they must have been regarded as items of value. There must have been some attraction of design or shape that led curious hands to pick them up and carry them along, to modify them for specific purposes, to touch them over and over.

Folklore from around the world offers insights into the motives of more recent collectors. One small trilobite species found in a province in China has been used as a medicinal "swallowing stone" for centuries. Some Welsh people still carry the ribbed rear portion of an Ordovician trilobite that is shaped like a pair of wings. These "petrified butterflies" have long been ascribed to an ancient spell of Merlin.

In the early 1900s a natural history buff named Frank Beckwith was digging in traditional Pahvant Ute territitory in west-central Utah when he uncovered a human skeleton. Within the rib cage lay a fossil trilobite. There was a hole drilled through the head of the trilobite, and its position inside the chest cavity indicated that it must have been worn as a pendant. A nearby mountain range contained an abundant deposit of this particular type of trilobite, which a Pahvant Ute acquaintance called by a name that Beckwith translated as "little water bug like stone house in." Upon inquiry, he learned that Ute elders used the fossils as cures for diptheria and sore throat, and wore them as amulets to afford protection in battle. At Beckwith's request, a young tribal member fashioned a necklace following a traditional design. When complete, it contained thirteen trilobite fossils, each drilled through the head and strung on a rawhide thong between hand-formed clay beads and tassels of horse hair.

— ·· ▰◈▰ ·· —

Back on the Tanglefoot, I turned over these stories along with the fossil in my palm, thinking of everyone who had touched those traveling trilobites—the Australian toolmaker, the wearers of the amulets, the Chinese physicians, the traveler on the Tanglefoot, the traders along the path, the schoolteacher, the scientists. I wondered how many of them had looked for more.

After a while, pulled by the lure of the search, I waded back into the snowmelt. The rushing water magnified the streambed into a swirling kaleidoscope. I plucked a random pebble and flipped it over. A mayfly nymph clung to the underside, its segmented body and bristling limbs echoing the trilobite form. The thrush sang on as I tested other rocks, pushing up riffles until my shins turned blue. But no more stone water bugs appeared in my hand.

A fresh burst of hailstones finally convinced me to call it quits. Shivering, I wiped the grit from my eel-white feet and re-laced my boots. On the trek back to my car, the single trilobite in my pocket began to bother me with the way it knocked against my leg at every step, and I stopped to draw it out. In the tired afternoon, the fossil had faded—its color dulling toward gray, the details of its anatomy sinking back into the stone. The word *relic* is rooted in the Latin *relinquere*, "to let go," and I thought about that as I tossed my prize beside the path and continued walking.

Ten steps out, a fading whisper of "once upon a time" reached my ear, and I turned around to retrieve my trilobite from beneath a tangle of budding serviceberry. The grooves and lobes were still there, however faint. I burnished the stone's bumpy surface with my thumb, slipped it into a different pocket, and carried on.

CHAPTER TWO

Water Dogs

Adult blotched tiger salamander *(Abystoma tigrinum melanostictum)*,
after a sketch by George Suckley, 1855

Robert's Cap ~

ROBERT HAD BEEN RIGHT on the verge of trouble all week. A short sinewy boy, lined in the face beyond any of the other seventh-graders, he possessed the kind of energy that kept him wriggling in his desk all day long. I was a guest teacher at his rural school for a unit on natural history, and although Robert made no secret of his distaste for books, he was happy to talk about anything remotely connected with hunting or fishing. But whenever I cut him off to present an assignment, he would pull his black baseball cap tight over his eyes and slide down in his seat for a sulk that generally lasted past the bell. That pattern held until Friday morning, when we ventured outside for a field trip.

A cool spring fog hung over the river as the class ambled down-stream toward the mouth of a small creek where, two centuries before, a group of Canadian fur traders had met an encampment of local Salish people. It was a comfortable walk, following long skeins of standing water bordered with many of the same berry bushes and wildflowers those earlier denizens would have known. We passed blooming camas lilies, and one of the tribal girls described digging their roots with her grandmother. We stopped to watch a cormorant make an underwater dive, and while the rest of us waited for it to resurface, Robert darted up and down the embankment, scratching through the grass like a buck rabbit. I was telling the class how the early traders sometimes ate the fat black birds for dinner when Robert motored up behind me, driving a bat-tered steering wheel he had pulled from the weeds. "Hey," he crowed, to the delight of his audience. "Think those mountain men left this behind?"

Beyond the trestle that spanned the creek mouth, we reached an expanse of floodplain. As their regular teacher and I herded the class through a gap in the fence, Robert veered off to peer into the opening of a large concrete culvert. "He'll catch up," his teacher assured me, and we pressed on.

The rest of the students had fanned across the grassland before Robert reappeared, his hands cupped in front of his chest as if carrying something fragile. The magnetism of discovery quickly drew his classmates back from the arc of the floodplain. Robert held his ground as they pressed around him, opening his hands to offer teasing glimpses of what appeared to be a dark-colored extra finger.

"It's a lizard," one boy announced.

"Don't touch it!" gasped another. "Those things squirt out poison from their skin!"

Robert did not say a word. He spread one palm flat so that everyone could see his prize, then curled his mud-stained fingers as pickets against the small animal's probes for escape. He drew a blade of grass down a wavy olive-green line that traced its sinuous spine.

"Thump its tail," a tall girl commanded. "It'll fly right off and dance beside the body."

I was opening my mouth to counter this flow of misinformation when a smaller girl shouldered her way through the circle. "That's no lizard," she said calmly. "That's a salamander."

She held out a fisted forearm, and Robert carefully placed the creature on her wrist. It began to walk, slowly but steadily, toward the back of her hand.

"Look how smooth its skin is," she instructed. "Anybody knows lizards have scales."

The girl rotated her wrist so that Robert's find stepped naturally into the protection of her palm. With her free hand she teased gumming bites from its harmless mouth, prompting a joke about a toothless grandparent.

"See, it won't hurt anything," she cooed. "I find these guys around our well house all the time."

One of the boys stepped forward for a closer look. "Hey, I saw one of those things poking around in the snow up on the mountain." He reached a tentative finger toward its head. "They're supposed to be cold-blooded, right?" he said, gingerly touching the tiny snout. "How can they do that?"

Other students had encountered the creatures as well. A girl confessed that she and her brother had found a pair of salamanders in a window well and decided to keep them as pets; they installed them in a glass casserole dish, only to have them both disappear the first night. Months later they discovered one of them behind the sofa, perfectly mummified.

Several members of the class knew you couldn't beat salamanders when it came to fish bait. Water dogs, they called them. One kid described the proper way to hook them, tugging at his own lower lip. He had an uncle who kept a washtub full down in his basement. "His are the kind with feathers on their neck," he added proudly.

"Those are gills, stupid," broke in the girl who held the salamander. "That's because they're just babies."

From the corner of my eye, I had noticed Robert step back and fade from view during the early stages of the discussion. He had been out of sight for only a few minutes when he returned, clutching his baseball cap to his midsection. To his obvious satisfaction, the class quickly gathered back around him. This time he revealed a hat chock-full of writhing salamanders, with all shades of green amoebic stripes. An excited voice asked where he had uncovered such a bonanza.

"Oh," he replied, playing it cool. "Around."

When I started in on the wisdom of putting the fragile creatures back where he had found them, Robert cradled cap to belly, his black eyes burning with the twin fires of possession and purpose.

"Take them back?" he asked, incredulous. "I'm the one who found them." Robert hugged his cap fondly, and a small smile of satisfaction creased his lips. "Besides," he said, "me and my little buddies here got some fishing to do this afternoon."

＋ ✦ ＝◆＝ ✦ ＋

I'm not much of a fisherman, but I do like salamanders. From their slender builds and the green stripes down their backs, I had recognized Robert's finds as long-toed salamanders, a species that I have uncovered everywhere from alpine lakes in Montana to rain forests on the Oregon coast. Yet you seldom see one of these secretive creatures, much less a hatful. Long-toeds belong to the aptly named family of mole salamanders (Ambystomatidae), who spend most of their adult life in solitude, hidden in burrows and crannies. Late every winter, as the ground begins to thaw, some unknown signal calls these hermits away from their catacombs. A few males begin to move toward the body of still water—anything from a puddle to a lake—where they began their lives. In succeeding days and weeks, pulses of other males follow. Some take to the water, but most seek shelter beneath any available cover. With the patience of hermits, they await the arrival of their female counterparts.

The writhing mass in Robert's cap had told me that just such a spring congress must be afoot. As soon as school was over for the day, I went back to the mouth of the creek. I knelt beside the dank culvert and began gently lifting rocks and rotting branches. Within minutes, I had uncovered a small selection of long-toed salamanders. I picked one up, wondering if this might be the evening when the first females trickled onto the scene and drew the waiting males into the water. These are creatures of the night, and their annual courtship rites are seldom seen by humans. Witnesses describe ponds roiled by the frenzied pummeling of competing males, followed by the undulating courtship dances of mating couples.

I looked at the animal resting in my palm. It raised its head very slowly, as if surfacing from underwater. White stars glistened from its moist, inky flanks. Its head wavered momentarily, then

13

bounced up and down. The salamander lifted a forelimb and spread four toes, each as fine as a stem of newly sprouted lettuce. The primitive wrist waved lightly in the air, its tiny digits reaching back toward the very beginnings of life on land.

The earliest known fossils that can be linked to salamanders appear in Asia, in rocks from the Triassic period around two hundred million years ago. When a volcano erupted in northern China fifty million years later, at the height of what we think of as the dinosaur era, a flow of lava overran a body of water not much larger than Robert's puddle. Just as Mount Vesuvius captured the breadth of daily life in Pompei and Herculaneum, the Chinese eruption exquisitely preserved a cross section of aquatic life in one small pond. Within this microcosm lay bodies of about five hundred amphibians of all ages, from larvae to adults, whose skulls, limb proportions, soft tissue imprints, and unique fused wrists are remarkably similar to the skeletons of modern salamanders.

From these Asian beginnings, salamanders radiated onto every continent, specializing as they plodded across space and time. Icthyosaurs and pteranodons came and went, but salamanders crawled on. The mole salamander family apparently arose in North America around thirty million years ago; from a locus in the valley of Mexico, they have populated almost every available habitat across our continent. Geologic upheaval and climatic change have isolated populations, and a bewildering variety of species has emerged, but the changes are relatively subtle: Basic salamander design has remained pretty much the same since that volcanic eruption in China long ago. The creature in my hand was a living relic of that primordial past.

Windmill Pond ~

THE DECREPIT WINDMILL stood alone in the scablands of eastern Washington, surrounded by overgrazed rangeland. Its stubby tower rose only about twenty feet above the ground, and its direction vane hung limp behind a spokeless differential. Near the base of its ruined sucker pump, a slim ellipse of cattails indicated the presence of a viable spring, which had been scooped out to make a small pond. The Bureau of Land Management had recently built a fence around the waterhole to keep out livestock, and biologist Todd Thompson was interested in what creatures might be making use of it. Considering the spread of barren ground around the pond, it looked like a most unpromising place for amphibians. Todd looked around at the battered landscape and shook his head. "You never know till you take a look, though," he said.

A curlew called from the open prairie as we slid down the short embankment in our chest waders and began to work our way through the suctioning silt, sloshing cold water near the tops of our bibs. A tree frog sang from the cattails, drawing an interested nod from Todd. April winds had blanketed the pond's surface with a tangle of tumble mustard, and we began examining woody skeletons soaked green with algal scum. After several minutes I raised a stalk lined with individual opaque globes, spaced along the stick like small peeled grapes. Each globe held a round black yoke rimmed with white. With a quick glance, Todd confirmed that we were looking at the spawn of a tiger salamander, another member of the mole salamander family. Tigers range over much of temperate North America, but in the entire Northwest there is only one variety, the blotched tiger salamander, which occurs along the mid-Columbia and some of its drier tributaries.

Circling the pond, we found more egg-bearing branches than seemed possible for such a small area. "That's one thing about salamanders," Todd said. "They're always going to surprise you." He reached down and scooped up a tiny red shrimp. Todd has visited hundreds of pothole ponds in search of salamanders, beginning with a field trip when he was in fifth grade, and he remains eager to talk about their mysteries. "People've tried to correlate them to rainfall, pH, dissolved oxygen, and fish, but it's hard to say what makes the difference. There are places where I find them thick as this one year, and when I go back the next spring—nothing. You just never know what you're going to find."

As we approached the cattails at the shallow end of the pond, we found an entirely different sort of jelly mass attached to the flotsam. These egg packets were smooth and limp, like stockings hung on a clothesline, with noticeably smaller eggs scattered throughout. When Todd held a branch up to the light, we could see that each ball enclosed an elongated creature with a wobbly line down its back and a tiny nub protruding from each side of its neck. These eggs belonged to a long-toed salamander. "See what I mean?" Todd exclaimed. "You'll read in books that tigers are found in the sagebrush country, while long-toeds belong in wetter, cooler places." But here they were, sharing the same scabland pond. Todd said he saw it every now and then, especially around the edges of the Columbia Basin. He was curious to see what would happen in the little pond beneath the windmill as summer wore on.

⤚ ⚏ ⤙

When I returned to the windmill pond two weeks later, its surface had changed drastically. The algae had retreated to the edges, and all the tumble mustard seemed to have sunk to the bottom. After several passes in my waders, I couldn't find a single egg mass, nor were there any signs of swimming larvae. It was a situation that called for a dip net.

The first swirl of the net dragged up a big glop of pure mud that rolled off the black and white patterns of many backswimmers, leaving them to rattle around the edges of the mesh. After a few moments, other creatures began to separate themselves from the muck: small crustaceans, purple worms, and leeches that twisted like sensuous leaves. It took a while to see the salamander larvae, lying perfectly still, like small-caliber bullets embedded in the slime. Lots of them.

The first two hatchlings that I plucked from the mud sported tails that were little more than transparent fins. Small bushy gills sprouted from the sides of their necks, and developing organs were visible inside their clear swollen bellies. I thought, tentatively, that they might be little tigers. The next one I pulled out seemed smaller, with knobbed appendages in front of its gill slits that looked like the balancing poles used by tightrope walkers. According to Todd, the balancers were a sure indicator that this was a long-toed salamander. Trapped in the net, both kinds of larvae looked like creatures still in the process of being born; released back to the water, they proved swimmingly alive.

In mid-May I took my kids out and impressed them by netting several larvae with every muddy sweep. Both kinds of salamanders still looked very fishlike, except for obvious legs budding off the front quarters of their smooth bodies. Both had golden eyes always on the glare. The tigers had put on appreciable weight, and some of their heads had grown so broad that they resembled bullhead catfish. At dusk we watched several of the larger ones hanging in the water column, their luxuriant gills waving like palm fronds in a tropical breeze.

The life of a salamander larva is fraught with danger; the creatures that feast on them range from great blue herons to fish. But if there are no fish present—and many Columbia Basin ponds are either too small or too alkaline to support them—it is often the tiger larvae that represent the most voracious predators in the pond. Carnivorous tigers have been known to gobble up other amphibian eggs, larvae, and even adults of their long-toed cousins. And yet in potholes where both species occur, the two moles seem to break the rules of logical ecology by breeding at just about the same time and growing in the water together. Somehow, the smaller, less aggressive long-toed salamanders must avoid being eaten, because they remain common. One key adaptation appears to be their rate of change from larvae to adult.

By summer's solstice, long-toed salamanders seemed to be a thing of the past—it was a tiger's pond now. Three swipes of the net produced five slurping larvae the size and color of gherkin pickles. Since no more than a small fraction of these larvae could possibly survive the journey to adulthood, it didn't seem like any great disturbance to borrow one of them for a while. We chose the biggest and most active pickle from the bunch and placed it in the bucket we had brought along, plucked a wapato plant that was sprouting nearby for shade, and headed home. Our captive was still very much alive when we transferred it to the miniature habitat we had prepared in a terrarium on the back patio. Its color was now a pure jade green infused with calligraphic lines. Recognizable digits crowned each limb—four on the front legs, five on the rear. Its silken gills, three to a side, were fringed with black lace and flowed like samurai decorations. Milky lips defined an outlandishly large mouth. We tucked the succulent wapato tuber into a patch of gravel in one corner of the tank and added a big scoop of mud from the pond to hold it down.

By the next morning the arrowhead leaves of the wapato had uncurled in glistening green, and a couple of its three-petaled flowers had burst into bloom. Below them mosquito wrigglers, a water scorpion, several striders, and multiple backswimmers were all carrying on as if they had never left the pond. The salamander, however, did not look so good. It seemed to be in shock, lolling and tilting in the water. Its belly was alarmingly distended. In the face of sudden movement, it would flail its roly-poly self down and out of harm's way, then bob awkwardly back to the surface. We peered helplessly into the tank until I remembered a woman who had told me about helping her dad catch salamanders for bait when she was a little girl. She said he always made her ride in the back of the pickup on the way home, keeping the pail that held the day's catch upright as they bounced toward town on rough dirt roads. Knowing that the larvae could gulp the sloshing water and choke to death, he taught her how to pick up any that appeared to be in trouble and use her fingers to massage their bloated bellies. She became an expert at burping them, laughing every time one expelled a mix of air and water with an audible bark—real water dogs.

Thinking of those swollen white bellies, I ladled our sick larva out of the tank and massaged its underside with my forefinger. Sure enough, a sharp burble escaped from its mouth. When I lowered the patient back into the water, it swam smoothly into the wapato leaves. We shooed the cat away and sat down in front of the glass to watch what might happen next.

Nosh'-Nosh~

MOLE SALAMANDERS, secretive though they may be, do occasionally appear among the oral and written records of the Columbia Basin. In the early 1900s, a Yakama elder told a story about Coyote journeying up the Teanaway River on the east slope of the Cascades. When Coyote came to a certain lake, he saw that the water was bad, and he decreed: "No salmon will come to this lake. Only *nosh'-nosh* will be here." Coyote returned downstream and built a waterfall to stop the fish, and from that day on, only nosh'-nosh, the water dog, lived in the lake. There he grew to great size. The elder explained that these water dogs belonged to the salamander family, and added that they were never used as food by his people.

Tribes around the rim of the basin, including Cayuse, Walla Walla, Nez Perce, Spokane, Kalispel, Flathead, and Kootenai, all have words for salamander. Like the Yakama, these tribes never utilized the water dogs for food, but several do associate salamanders with the idea of bad or dangerous medicine. This could be attributed to the animal's mysterious habits and confounding life changes, and such ideas are by no means confined to Native Americans. In European lore, salamanders spontaneously generate themselves from the flames of a household hearth, and their parts often figure in recipes for witch's brew. In Japan, the word *ryuu* means both "salamander" and "dragon."

The reaction of the Scottish botanist David Douglas was similarly ambiguous in the midsummer of 1826, when he followed a tribal trail that wound between scabland coulees and the Palouse Hills of eastern Washington, through "an undulating woodless country of good soil, but not well watered." Douglas enjoyed the day's ride with his usual fervor for new places, but his enthusiasm was somewhat dampened at suppertime: "We were obliged to cook from stagnant pools full of lizards, frogs, water snakes." Many people, past and present, call any small four-legged animal of a certain shape a lizard. But since true lizards don't swim, salamander larvae are the only creatures that really fit Douglas's description.

Thirty years later, naturalist George Suckley made a beautiful drawing of a tiger salamander while surveying a railroad route along the Columbia, but apparently no scientist probed their larger range until U.S. Army surgeon Basil Norris paid a visit to the northern edge of the Palouse in early June 1886. During an investigation of the purported alkaline healing properties of Medical Lake just outside Spokane, Dr. Norris captured a couple of peculiar "reptiles, the species of which has caused so much controversy in a local way for years." Seeking an authoritative opinion, he shipped the swimmers east to the Smithsonian, and a few weeks later he received a reply from its esteemed director, Spencer F. Baird.

Dear Doctor,

The specimen referred to in your letter of June 12th was duly received, and, on an examination, proves to be the larva, or immature stage of the salamander. It is one of the so-called water lizards, found in wet places, under logs and stones. We are very glad to get the specimen as it is considerably out of any range known to us. We should like to have more of these creatures as they are probably quite abundant in your neighborhood.

James Slater, a Tacoma college professor and salamander buff, paid a visit to the source of this early specimen in September 1930. In the town of Medical Lake he spent an afternoon searching for the local water lizards in vain. Looking for inside information, Slater spoke with a young man at the swimming beach, who promised that he and his friends could supply plenty of the "dog-fish" (meaning "fish with legs") after dark. Sure enough, a little after eight a few local men gathered and kindled a bonfire before stepping into the lake to drag a seine net. To Slater's delight, their pass captured a dozen larval salamanders.

As soon as that crew left, another group appeared. Slater learned that since July these men had been driving from Spokane and catching salamanders to sell as fish bait. "I suppose we should call them salamandermen instead of fishermen," he wrote. While the professor pondered whether the creatures might be attracted by the light of the bonfire, the seiners brought fifty-five good-sized larvae ashore. The catch included two adults with the distinct dark and light pattern of the blotched tiger salamander. Slater made sure he got that pair for himself, and accepted a few of the larvae as well.

The leader of the Spokane seiners told Slater that year after year, colored animals started coming up in the net around August 10 and continued to appear until the season ended around mid-September. His personal record for salamanders taken was 159 in a single pass of the net, and 209 dozen in an evening. The creatures caught that night varied in length from three to seven inches, which he deemed about average. The salamanderman could tell that his quarry's abundance was tapering off, and he figured this would be his last trip of the year. Before departing, he confided to Slater that the going price for water dogs at Spokane bait shops was fifty cents a dozen—not a bad take in the midst of the Great Depression.

Sea Change ~

AS WE DRIFTED THROUGH the dog days of summer, change was afoot in our terrarium. The wapato shed its white petals one after another, and the sepals formed round green seed pods. Our salamander larva took to lying on the surface of the water at dawn and dusk. It would ride at the level of the tangled weeds, then sink a bit, pushing away with soles and palms turned outward as if the water were a supportive wall. Sometimes it would stretch out all eighteen of its toes, with one digit on each side breaking the surface. Its eyes began to bulge from its head, growing from flattened inset disks into round buttons. Odd swellings appeared along both sides of its neck, and its gills began to shrink from the feather boas of their prime. Its body developed distinct dark patches that dripped into parallel bars, but the belly remained clear white, bordered by a beautiful pattern of black stipples. Sometimes it would make a snap that might have been feeding. Occasionally it would burp out an air bubble with the sound of an old man spouting a good stream of tobacco juice, as if it might be learning how to breathe. But most of the time it hung still, showing grave indifference to the activity that whirled around it.

Then came a day when we found our captive lying on the surface, completely motionless, supported only by plant fibers. At first the kids were sure it was dead, but they misted it with a spray bottle over and over until, with excruciating slowness, the patient swam to the far end of the tank and rested its head and shoulders on a flat rock just clear of the water. To our astonishment, we could see that its entire front end had assumed the eerie, varnished sheen of an Andean mummy. For the next several hours, it did not move one iota. In the cool of the evening, the larva slowly lifted its head. It was then we realized that we could no longer see its gills. The muscles along the sides of its neck flexed, and the gill slits pulsated visibly, but those outrageous feathers, for so long our larva's most visible feature, had disappeared. I had read about amphibians resorbing their gills during metamorphosis, but nothing had prepared me for the fact that an appendage half as long as the animal's body would disappear into its neck.

The salamander hung in limbo between infancy and adulthood, between life and death, between the worlds of water and land. Now nascent lungs had to inflate with small gulps of oxygen not just occasionally, but with a continuous rhythm. The membrane of skin had to make the switch from water to air. Limbs accustomed to swimming had to assume the posture of a tetrapod; a body made for floating had to comprehend gravity. The creature was undergoing a metamorphosis that defined its whole existence, a monumental event that reprised not only the life history of its species, but that of all amphibians, and of Earth itself.

The salamander still lay in a light coma when night fell, and the next morning it was nowhere to be seen. We searched for many anxious moments before spotting the tip of a tail peeking out from under a spruce bough in the dry part of the tank. When we lifted the branch, we found ourselves looking at a completely transformed creature. Its head, broad, smooth, and smiling, seemed to have expanded, while its body had shrunk as if tightly wound in plastic wrap. The phoenix rocked its big head forward and back. Its neck throbbed with slow but steady breaths. Fore and hind legs moved once, then again, very slowly.

Every morning for the next several days we found it in a different place, squeezed into a rock crevice or tucked beneath a slice of bark. Sometimes it flopped into its little pool and swam turtle style, matching strokes with arms and legs of opposite sides. In the light its skin glowed like the oiled parchment of an antique map, with sharply defined islands of mustard and ebony. Its tail assumed an elegant taper, and fleshy doughnuts surrounded those periscope eyes. My ten-year-old brought an earthworm from the garden and waved it in front of the salamander's nose. Its head ratcheted up one cog, then another, then lunged forward and seized the prey. Taking a ritual bow, the salamander dropped its head and shook the victim with a single violent snap. It took several minutes for the two dangling ends of earthworm to disappear, with periodic gulps, into the soft crescent mouth.

At the end of August, after eight bone-dry weeks, a morning thundershower rolled across the scene, and raindrops pelted our desiccated world. Within moments the salamander had ascended to the highest tip of the spruce bough that decorated the terrarium. Its head wobbled back and forth with every new drop from the sky. One eye blinked. It was feeling air and moisture together, an animal made for rain. As succeeding nights grew cooler, I kept imagining all those larvae back in the windmill pond, now transformed into adults and preparing to leave the water to find a secure burrow or crevice for the winter. We decided it was time to return our captive to the wild.

❄❖❄

Dust enveloped the car as we pulled up to the ragged windmill, leaving us to wonder once again how a creature that required moisture could survive in such a dry place. The pond had shrunk to a fraction of its summer size, and across its reduced surface, brown wapato leaves were covered with black dots of insect frass. Green tree frogs were still hopping all over the plants, but scoop after scoop of mud failed to bring up any salamanders. Then on one of the last sweeps, a familiar shape snaked through the net. It proved to be a beefy tiger larva at least six inches long and very broad in the head. Its gills were huge, and its legs were strong and flailing, but the eyes still lay flat, which lent it a mean, threatening look. It was a neotene.

The hormones that trigger metamorphosis do not always flow at the same time for all the salamanders in a pond, and some larvae may not transform for a year or even more. In certain cases, such creatures can reach sexual maturity without ever leaving the water, a state of retarded development known as neoteny. These morphs, which can grow to outlandish size, often act like monsters in the pond, preying on their own kind. It is sneaker-sized neotenes, flailing in the mud of disappearing ponds, that leave sageland farmers sputtering with cries of "walking catfish!"

The first known written mention of mole salamander neotenes came from the Aztec capital of Tenochtitlan, where sixteenth-century Franciscan monks traced stone carvings depicting a god named Xolotl. This deity bristled with extra body parts, especially odd numbers of fingers and toes, and it appeared to sprout layers of feathers from the back of its neck. When the Franciscans inquired into the meaning of the name Xolotl, native responses included water slave, water servant, water sprite, water monstrosity, water twin, or, most familiarly, water dog. Brother Bernardino de Sahagun, assigned to teach a group of Aztec youths, learned from his students that Xolotl was closely associated with the *ajolote*, an aquatic form of salamander that thrived in the necklace of canals and lakes that embraced Tenochtitlan. "Like the lizard, it has legs," the boys told Bernardino. "It has a tail, a wide tail. It is large-mouthed, bearded."

The students showed their teacher the strange gilled creatures, some up to a foot long, and explained that they provided an important food source in waters that supported few fish. "It is glistening, well-fleshed, heavily fleshed, meaty. It is boneless—not very bony; good, fine, edible, savory: it is what one deserves." When, after forty years of labor, Brother Sahagun published his landmark account of Aztec culture and natural history known as the Florentine Codex, he included an entry with the title "Axolotl." The accompanying illustration depicted a creature with four legs and flowing gills, accurately representing a creature exactly like the neotene in my net.

I let the big pond monster slither away, then returned to the car and fetched the bucket that held our much smaller, newly metamorphosed adult salamander. We walked around the pond to size up the situation. An area of cracked mud was crisscrossed with the tracks of coyote and badger, skunk and raccoon, and the three-toed prints of ravens, gulls, and herons. Any salamander that ventured out on this hardpan would be dancing at a predator's ball. Across the way we spotted a badger burrow, and around from that a bank so steep we couldn't imagine any salamander making the climb. But down on the cattail end, a nice pile of drain rock rested in a damp seep. The rocks were of different sizes, with plenty of gaps and crannies where a little animal could hide. That was the place we felt our little tiger salamander deserved; that was where we tipped the bucket and let our captive go.

CHAPTER THREE

White Shield

Sagebrush sheepmoth *(Hemileuca hera)*

Flight ~

WITH NO BREEZE TO STIR the air and no sign of rain for weeks, the August morning was heating up fast. As open bunchgrass began to crackle under the sun, the whole landscape seemed to slide into the protective arms of the nearest plant. One little sagebrush lizard, colored like its namesake's leaves and flowers, positioned itself on an outside branch to catch some rays. A thrasher sailed in to touch a crown of rabbitbush, then departed without a sound. Beneath the crinkled leaves of a balsamroot, the bright red underwings of a captured grasshopper shone through the open curtain of an orb spider's web.

Walking a fallen fenceline, I spied what appeared to be a small white shield shimmering in and out of focus near the top of an ancient sage. From a distance, I took it to be a picked vole carcass or a cricket shell—the remains of a meal pinned aloft by some efficient shrike. But as I drew closer I saw that it was a motionless insect. Black eyespots blinked from the center of each wing, and black half-diamonds along their margins pointed straight at the open orbs. It was a female sagebrush sheepmoth, newly emerged from her pupa beneath the ground. Life's liquid had inflated her wings, and she had scrambled up the branches to assume her position on the very top sprig of the bush.

I stepped cautiously forward, but she did not fly. Fine golden-orange veins threaded across her black and white wings, precious metal and fruit combined. The eyespots of her forewings came into focus as calligraphic black Cs, while those on the hindwings dripped into the shapes of fancy sixes and nines. The moth's lower body, patterned with alternating bands of gold and black, could have belonged to a fecund bumblebee, and she accentuated the similarity by slowly curling the tip of her abdomen like a bee intent on stinging. I brought my face nearer, until I could see how her shoulders were wrapped in a luxurious stole woven from furry scales of foxy sorrel. Rich auburn tones topped her pate and swept to the tips of her wiry antennae. Short combs grew off each antenna segment, and a few on the outside of the left one looked bent or damaged.

A puff of wind sprang up, and the moth adjusted her legs for a better grip on the sagebrush. No matter how closely I approached, she did not budge. She pulsed her abdomen again, and I guessed she was releasing a plume of pheromone to ride the breeze. Somewhere in the acres of sagebrush that surrounded her, flying males were waving their own leafy antennae, equipped with elaborately combed receptors sensitive enough to pick up a female's perfume from miles away. Like flicking darts, they were careening across the top of the sagebrush in wild zigzags, neither slowing to rest nor dipping for nectar. A male will quarter the wind until he intersects a scent plume, then row against the current of the breeze as he gauges the pheromone's concentration against the lay of the land. The closer he approaches to the source, the more directly into the wind he flies, homing in on the sole purpose of his adult life. He has only a few days to fulfill his mission, for adult sheep-moths have no mouth parts for eating, no way to refuel. The perched female usually accepts the first male that finds her; soon after her eggs are fertilized, she flies off in search of a suitable host plant. For the rest of her brief life, she circles the lower branches of a succession of sage bushes, laying rings of pearly sage-colored eggs.

⊶⊷ ▰◆▰ ⊶⊷

Early last September, on a gray morning with a cold wind bearing down, I came upon one such female. She was lying stock still on the ground near the central trunk of a spreading sage, hemmed in by the forest of branches around her. Her powdery scales had flaked away so that she looked more like a piece of littered newspaper than a fine white shield. Sharp twigs or a pecking bird had tattered both hindwings, and her antennae were folded horizontally across her beady eyes. The spot high on her shoulders where burnished chestnut scales once flashed was now rubbed bare.

And yet the gold filigree veins trailing through her wings still glowed with life. As I watched, she began to move her forewings very slowly up and down. After some minutes the hindwings also began to quiver; the shivering spread at an almost imperceptible pace until her entire body shook like some early aircraft warming its engine for liftoff. Both antennae stretched out, buffeted by the wind, and the moth began to crawl. Her legs clutched a low branch and hoisted her body off the sand. She fell back, scrabbled to gain the branch again, then teetered out to its end. There she hopped clear of the entangling twigs and lifted off. Within seconds she was up and away, flying the same crazy zigzag as the males, seeking another gnarled trunk on which to set out one more ring of eggs while she still possessed the strength.

Summer Nights ~

SHEEPMOTHS BELONG to the Saturniid family, giant silk moths revered for size, pattern, color, and startling eyespots. Growing up in the Carolinas, my sisters and I would sometimes find a silk moth clinging to the screen door on a humid summer's eve, or dripping from a tree limb after a thunderstorm. We would run to fetch our mother, and she would stop whatever she was doing and follow us outside to shimmer in the visitor's presence. But Mother was not content merely to admire the moth; she wanted no less than to possess it.

My job at this point was to grab a Mason jar from the kitchen while she fetched a cotton ball and her red can of spot remover. My sisters and I would watch wide-eyed as she primed the cotton with a quick tilt of the can, gave the ball a wave to evaporate any excess liquid, then dropped it into the jar. While I held the lid on tight to prevent any fumes from escaping, she would pluck the moth from its perch, expertly fold its wings together, and gently place the insect inside our killing jar. For a moment, her bony wrist would hover over the jar's wide mouth, then she would slide her hand away so that I, as lid man, could close the moth inside.

Mother insisted that we wait for exactly her prescribed time, and I can still see the red pentangle of the minute hand on her fancy cocktail watch creep from one bold black line to the next. With each tick, a new kaleidoscopic pattern would emerge from the center of the watch's face, golden triangles that grew to teardrops and held me mesmerized until the time was up. While the seconds crept past, she would peer into the jar, pointing to the royal purple line that traces the leading edge of a Luna moth's wings, then compare their peculiar shade of pale green to the languid summer moon. If it happened to be a Cecropia moth, she would show us its dark chestnut pattern and explain how Cecrops sprang from the earth. On the night we peeled a Polyphemus moth from a big red oak, she branded the single eye of the giant Cyclops on my mind forever.

Living among the lush hardwoods of the South, we never dreamed that wild silk moths might also flit through the high deserts of the American West. It never occurred to us that such a creature might fly by day instead of night, or take its name from a humble grazing animal rather than from a resplendent god. But if we had been living in Walla Walla instead of Waxhaw, Mother would have known about such things. On the hottest of August afternoons, she would have called for pillowcase, coat hanger, and broom pole to fashion one of her special nets, then sent us trundling through the sagebrush forest in search of *Hemileuca hera.*

When we returned victorious, our sheepmoth would have provided Mother with more fodder for her mothology. She would have spelled out the genus name and reminded us that *hemi* meant half. Pressing her palms together, she would have showed us how the moth's folded wings conjured exactly half of the special shield carried into battle by the Trojan soldier Leucaspis. She would have made us proud when Leucaspis used this badge so effectively that his companions dubbed him White Shield.

Leaping to the species name, she might have spun a yarn about the jealous Hera, who always knew when Zeus was up to mischief. The day she caught her husband with a snow-white heifer, Hera saw the maiden Io in its face, and demanded that the tosser of thunderbolts give her the cow. Zeus, of course, could not refuse. Hera the goddess tethered Io the heifer to a tree and called on Argus, her servant stippled with a hundred eyes, to guard the girl closely. To counter Hera's move, Zeus coaxed quicksilver-tongued Hermes to distract Argus. Hermes told stories long and slow until Argus's many eyes were nodding in and out of sleep, just like the blinking eyespots of a sheepmoth in flight.

No matter how carried away she got with her stories, Mother always stopped after precisely five minutes had ticked past on her watch. At her nod, I would carefully uncap the jar so she could touch the moth's body and pronounce it dead. Then, ever so gently, we would pin the outstretched wings on a piece of cardboard wrapped with black velvet. Our goal was to preserve the essence of the creature before its delicate body stiffened and the brilliant colors lost their sheen, and we took great pains to make it look as though the moth had simply settled onto our board for a short rest, ready to lift off again with the fall of night. We mounted many moths before I realized we would never come close.

The Collector ~

THOMAS NUTTALL FELL under the spell of the outdoors early in life in his native Yorkshire; in 1808, at the age of twenty-one, he gave up a printer's apprenticeship and sailed to Philadelphia. There, with the encouragement of a vibrant scientific community, he concentrated on plant studies up and down the Eastern Seaboard. In 1811, aiming to botanize farther afield, he took advantage of an opportunity to accompany the Pacific Fur Company's overland expedition up the Missouri River as far as the Mandan villages, in what is now North Dakota. As the troop made its way across the Great Plains, the greenhorn naturalist came into his own, collecting like a man possessed. Ground squirrels, birds, snakes and lizards, beetles, ammonite fossils, flashy minerals, and chunks of petrified wood steadily filled up boxes alongside his plant specimens. The French-Canadian boatmen made great sport of his habits, especially after they found that his gun was often plugged with dirt from being used as a digging stick or loaded with seeds to ensure their safekeeping. Thomas Nuttall chuckled at his own peculiarities and continued to revel in all things natural.

When he returned to the East Coast, he embarked on an intense quarter-century of research and teaching, producing a comprehensive manual of North American plant life and a similar project on the continent's birds. Even though his publications became early standards in their respective fields, Nuttall always regretted that he had never made it to the Rocky Mountains. Then, at forty-eight years of age, he was offered a chance to visit the Northwest by his friend Nathaniel Wyeth, a Boston businessman who was leading a trading venture to the Columbia. Nuttall's collecting instincts sprang back to life; he resigned his teaching post at Harvard and invited John Kirk Townsend, a Philadelphia physician and bird enthusiast half his age, to join the excursion.

Wyeth's expedition embarked from Independence, Missouri, in the spring of 1834. The procession included furmen, the two naturalists, three Methodist missionaries, and a herd of cattle. One of the missionaries, noticing that Nuttall often hurried ahead of the crowd to collect plants before they were trampled by the livestock, wrote that the botanist's "characteristic ardour in his favorite pursuit has not been lessened by the lapse of three and twenty years."

Long before they reached the Rockies, the naturalists had cast aside their extra clothing, shaving kits, and soap in order to make more room for specimens. The party crossed South Pass, then proceeded on to the Snake River. A few miles upstream from the mouth of the Portneuf River in what is now southeastern Idaho, Wyeth stopped to construct his Fort Hall trading post. The delay provided a chance for John Townsend to experience the thrill of running buffalo, but Nuttall, wary of Blackfeet raiders, stuck close to camp. During the three weeks it took to build the post, he snipped samples of various willows that grew along the riverbanks and watched western marsh wrens scurry among the rushes.

The Americans left Fort Hall on August 6, crossing the Snake to find themselves engulfed in a wide sandy plain scoured by hot winds and covered with "luxuriant sage bushes." Their time at Fort Hall and their journey across the Snake River Plain coincided perfectly with the annual mating flights of two species of sheepmoths. One was a crisp black and white, while the other sported melon hues, which varied from Crenshaw to cantaloupe. Thomas Nuttall captured and pressed several individuals of both types, perhaps to demonstrate the range of their coloration and to give himself a better chance of getting home with an example of each. As with so many of the small flora and fauna the naturalists saw along their journey, these moths were not yet known to the halls of science. At that moment in time, the chances of their delicate wings making it from the far reaches of the Snake back to the classifying rooms of Philadelphia must have seemed frightfully slim.

After a difficult four-week journey over shattered basalt fields and sideways mountain ranges, Wyeth's party reached the Columbia River at Fort Walla Walla and continued downstream on horseback to the Dalles. There they procured three canoes to float down to Fort Vancouver, but they had been on the water barely an hour when an afternoon gale swamped the boats. As they waited through the following day for the wind to die down,

Townsend, whose bird skins had escaped damage, watched his colleague attend to his precious specimens.

> Mr. N.'s large and beautiful collection of new and rare plants was considerably injured by the wetting it received; he has been constantly engaged since we landed yesterday, in opening and drying them. In this task he exhibits a degree of patience and perseverance which is truly astonishing; sitting on the ground, and steaming over the enormous fire, for hours together, drying the papers, and re-arranging the whole collection, specimen by specimen, while the great drops of perspiration roll unheeded from his brow.

The sheepmoths Mr. N. collected along the Snake River must have required some of his most tedious attention.

At Fort Vancouver, Nuttall and Townsend boxed their various collections and shipped them by boat around Cape Horn to Philadelphia. A year later, Nuttall followed the same route home. His vessel met rough weather trying to round the Horn and was twice beaten back into the open Pacific. On the third attempt a dense fog settled around the ship, confusing her course for days. The clouds finally lifted to reveal a familiar landmark that proved she had made the turn into the Atlantic. Seaman Richard Henry Dana, who had been a student of Nuttall at Harvard, never forgot either the drama of the moment or the reaction of his professor to the news: "Even Mr. Nuttall, the passenger, who had kept in his shell for nearly a month, and hardly been seen by anybody . . . came out like a butterfly, and was hopping round as bright as a bird."

Nuttall sailed into Boston just as John James Audubon arrived in the city to sell subscriptions for the fourth volume of his monumental *Birds of America*. Audubon, who had never visited the West, was anxious to obtain study skins and background elements from those regions. He breakfasted with Nuttall and a few weeks later the two met again at the Academy of Natural Science in Philadelphia, where Nuttall was sorting through his barrels of treasure from the Columbia River. The collector supplied the artist with birds' nests, specimens of western plants, and pressed insects. Audubon purchased duplicates of more than ninety bird skins sent back by Townsend and took this bounty south to the home of his friend John Bachman in Charleston, South Carolina, where he painted new figures for his opus. Among them was a composition depicting a pair each of Say's phoebes and western kingbirds, two totem birds of the arid West. The phoebes are perched on a leafless branch, their bills pointed up to worry a buzzing fly. The two kingbirds, joined by a scissor-tailed flycatcher, focus their attention on a pair of moths.

The patterns of these two moths are almost identical. Single eyespots, slightly crinkled, stare from the forewings of each insect; both heads sprout orange hairs; both abdomens are graced with similar dark and light bands. Only the colors of their wings are different: Audubon painted one moth's wings a creamy white, while he dabbed the other's with the pink of the phoebe's belly in front and the yellow of the kingbird along its trailing edges. These are Thomas Nuttall's Snake River sheepmoths, appearing in the public eye for the first time. But while the birds' bodies carry the animation of life, the moths look as if they were hammered flat on the canvas. Audubon, painting from pressed carcasses two years old and two thousand miles removed from their home, had no more hope of capturing the living magic of those Saturniid moths than my mother and I did when we pulled them, oily and stiff with spot remover, from our killing jar.

His portraits completed, Audubon left the two crusty moths in Charleston with Bachman, who sent them to a friend in England, who decided they must be the male and female of a new species, which he assigned to the goddess Hera. Meanwhile, back in Philadelphia, Nuttall had presented several more sheepmoth specimens to the painter and naturalist Titian Ramsay Peale, who made etchings of a pair for his proposed *Lepidoptera Americana* and named them in honor of their collector. Eventually the moths were recognized as separate species, and the taxonomical confusion was resolved; the black and white moth retains the name of Hera, while the melon-colored one honors Professor Nuttall. The desiccated remains of both sheepmoths remain forever pinned onto Plate 359 of Audubon's *Birds of America*, with those western kingbirds eying them as if they might make one last meal.

Sage Dwellers ~

SINCE MOTHS HAVE no hard parts, their fossil history is very sketchy, but some lepidopterists believe the ancestors of western sheepmoths may have fluttered across the Bering land bridge at the very beginning of the last great Ice Age, more than a million years ago. Pollen counts from ancient lake bottoms reveal that most of the dryland shrubs we now associate with the intermountain West have been around since that time, following the ebb and flow of glacial movement. Insects of all sorts would have lived in concert with sagebrush and related shrubs like antelope bitterbrush, saltbush, and greasewood. Sheepmoths would have been a part of this scene, coevolving with the plants; over time different species developed to mirror the subtle diversity of their environment. Hera's variety settled on big or tall sagebrush as its primary food plant, while Nuttall's came to depend on the sprigs of antelope bitterbrush. In the millennia when these shrubs blanketed the arid West, both moths would have thrived, and it is easy to imagine clouds of white and melon-colored males roving across the steppe in search of females on clear, windless, late summer days.

The years since the sweep of Thomas Nuttall's net have brought great changes to the sagebrush country. Grazing livestock, invasive weeds, large-scale agriculture, a complete shift in the pattern of water transport, and human sprawl have flooded through continuous stands of sage and bitterbrush. In the face of such alterations, sheepmoths have retreated to islands of healthy shrub-steppe that hearken back to what must have been a fluttering heyday for *Hemileuca*.

Professor Nuttall left behind no field description of sheepmoth males performing their aerial displays, or expectant females blossoming on the tops of their food plants. Most of the other early travelers to the sagebrush plains of the Snake and Columbia Rivers were so anxious to reach the verdant coast that they paid little attention to flying insects. While tribes of the southern Great Basin were keenly aware of the life cycle of such moths—figures on Miembres pottery depict spiny-backed caterpillars, and Paiutes around Mono Lake trapped the grubs of a close sheepmoth relative in pits, then roasted their bristles off in sand and dried them for use in winter soups and stews—Plateau peoples seem to have no stories or words directly connected to the sheepmoths. Perhaps it was because the abundant fish in this region provided bigger packages of more easily obtainable protein. Perhaps as the moths decreased in number, they simply faded from memory. But just because sheepmoths cruise beneath the reach of human radar doesn't mean that some aren't still out there, making their rounds.

In early spring, as the fresh new leaves of dryland shrubs begin to burst forth, the tiny egg clumps laid by female sheepmoths the previous summer finally hatch, and the larvae congregate into fuzzy black lumps on the branches of their natal bush. Close inspection of one of these writhing clumps reveals a knot of dark caterpillars pulsating with the rhythm of a single organism. Individuals crane their heads out of the pack, wave their six thoracic legs, then disappear back into the clump as others rise around them. This gregarious behavior allows the larvae to absorb more solar energy during the day and to keep warm on cool nights; it may also be a defense posture meant to startle preying birds and fend off insects. Amidst all the twisting and turning, the dozens of larvae inside the bundle are busy chewing away on tender leaves, fueling a growth spurt that will carry them through a succession of developmental stages called instars.

Over the next few weeks, the larvae increase remarkably in size, and the clumps separate into single caterpillars that pull their way along branches on five pairs of knobby prolegs. With supple strength they balance on their hind ends while twisting and flexing in acrobatic displays. By the time it reaches its fourth instar, a sheepmoth caterpillar is about the size of a baby's finger, with yellowish seams running the length of its midnight body. Whorls of stiff yellow bristles, like tiny shaving brushes, sprout in tidy rows along its back and sides. These little spines contain a stinging chemical that can raise welts in the mouths of grazing livestock or on the sensitive skin of curious humans. Despite such defenses, their mortality rate is high; many larvae fall to hawking birds or play host to the eggs of parasitic flies and wasps.

As spring's freshness dries to summer's heat, the surviving caterpillars give up their voracious habits and slither down their bush to touch the earth. Like dogs settling for a nap, they circle the base or wander to adjacent bushes before slowing to a halt. Their bodies undulating from head to tail, they burrow into the thin desert soil. Then, instead of spinning cocoons like most of their silk moth kin, they exude a liquid gel that hardens into a glassy capsule. At first the new pupae glow like self-made amber, but air and time soon darken them to a glossy black.

Inside these fragile chambers, the caterpillars undergo a profound transfiguration. The larval mouths disappear, and complex eyes pop to the surface of the head. Segmented antennae develop, and dormant cells form four fragile, amorphous membranes that will one day become wings.

When the transformation is almost complete, subtle forces determine the timing of release. In the Columbia Basin, where warm weather lingers, most receive the signal to emerge at the end of summer, and so spend only one season underground. The biological clocks of others may keep ticking through an entire year, or even two, so that some sheepmoths do not emerge as adults until twenty-seven months after their descent. Only then do they wrench free from their chrysalises and dig upward through the soil. In the open, they pump elixir through the network of veins that gives shape to their new appendages, then begin the long climb to the top of their mother sage, where they spread their wings to match Leucaspis's fine white shield.

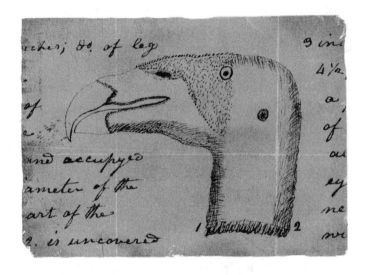

CHAPTER FOUR

The Beautiful
Buzzard of the
Columbia

Head of a vulture from Meriwether Lewis's journal, February 17, 1806

Taken Alive~

DURING THE LAST WEEK of October 1805, the members of the Lewis and Clark expedition struggled around the complex portage of the Dalles and began their descent into the Columbia Gorge. As the landscape changed from desert to rain forest, fall migration brought myriad swans and skeins of white cranes flapping overhead. The 30th dawned rainy and foggy, and around midday the explorers halted for a bite to eat in a timbered bottom at the mouth of the Wind River, just upstream from the Cascade rapids. In his daybook Clark noted that a soaring bird had graced the scene: "saw the large Buzard white head and part of the wings white." At first reading, this bird sounds like an osprey or an immature bald eagle; in those days, *buzzard* was a catchall term for any raptor, and both of these fish-eaters show white patches on their heads and underwings.

The bird also caught the attention of Sergeants John Ordway and Joseph Whitehouse, who both mentioned seeing turkey buzzards with white feathers under their wings. When Clark sat down that evening to flesh out the day's events, he included additional information: "this day we Saw Some fiew of the large Buzzard Capt Lewis Shot at one, those Buzzards are much larger than any of their Spec[ies] or the largest Eagle. White under part of their wings &c." Now it is not so easy to identify the birds. What raptor is larger than an eagle? Certainly not turkey vultures, which in any case usually migrate south from the Columbia country in early fall. But their pale heads and underwings can appear whitish in a mist, and perhaps a combination of the rainy weather and dense forest simply confused the men's perspective on a lingering flock of common turkey buzzards.

After their encounter with the large birds and a final portage around the Cascades, the crew floated on to the mouth of the Columbia. In late November, Clark and ten of the men were on a trek to Cape Disappointment on the north side of the river when they came upon the carcass of a dead whale. Some of the same large buzzards had already discovered it, and hunter Reuben Fields wasted no time in bagging one. With the strange bird now in hand, Clark calculated its weight as twenty-five pounds. Its outstretched wings measured nine feet, six inches—three feet greater than even a good-sized turkey vulture. The end of the tail stretched three feet, ten and a quarter inches from the point of the bill. A single flight feather trailed out for two and a half feet. The middle toe, five and a half inches long, carried an inch-long nail. From the sum of their training and experience, the captains had no knowledge of a bird that fit these outlandish dimensions.

In mid-February, George Shannon and Francis Labiesh went out hunting from Fort Clatsop on the south side of the river and returned with a wounded buzzard that was still alive. Without commenting on how the pair managed to subdue and transport a creature that could have wrapped its wings around both of them, Meriwether Lewis set to work analyzing the captive in the systematic manner he had been taught by Thomas Jefferson. Lewis peered into the buzzard's dark eyes to discern a sea-green pupil floating in a pale scarlet iris. The yellowish orange of the bird's bare head shaded toward a fleshy pink along its neck. The white stripe under the wing measured about two inches in width, and its other feathers shone glossy black. The short, whitish legs were very rough, and feathered only to the knee, and the blunt nails seemed quite different from the talons of a hawk or an eagle.

Labiesh said that when he approached the wounded bird, it made a sound very much like the barking of a dog. Lewis, always interested in a creature's vocal abilities, pried open the vulture's beak as if searching for the source of its bark. He found a firm, broad tongue, which folded upward into a longitudinal groove, its edges "armed with firm cartelaginous prickkles pointed and bending inwards." While Lewis probed the buzzard's mouth, William

Clark sketched its head on the edge of a map he was working on. More refined drawings appear in both captains' journals, surrounded by descriptions of the bird. The sketches depict a bare head with a circular ear opening below the line of the eye, and the beginnings of a feather ruff midway down the bird's naked neck. If their measurements and written descriptions were not sufficient, these drawings leave no doubt that the explorers were looking at the bird we now know as the California condor. "I believe this to be the largest Bird of North America," wrote Lewis, who had seen a fair selection of the continent's big birds.

Full of fresh observations, Lewis expounded on the bird's behavior. "This bird fly's very clumsily, nor do I know whether it ever seizes it's prey alive, but am induced to believe it does not. we have seen it feeding on the remains of the whale and other fish which have been thrown up by the waves on the sea coast . . . but I have no doubt that they also feed on flesh." Lewis was essentially correct: Their short talons and relatively small beaks prevent condors from taking live prey. Like other carrion eaters, their habit is to land on the ground some distance from a prospective meal, then approach in cautious circles. They are not particular about their food, as long as it is already dead.

When time allowed during the winter months, the captains tracked the avian life around Fort Clatsop, writing that "a small Crow, the blue crested Corvus and the smaller corvus with the white breast, the little brown wren, a large brown sparrow, the bald eagle, and the beatifull Buzzard of the columbia still continue with us." Certainly no one today would lump the mythical condor with the Northwest crows, scrub and Steller's jays, winter wrens, and brown towhees of a coastal community, but in the winter of 1805–06, the orange-headed scavengers formed part of the local scene. The buzzards were still around in early spring as well; in mid-March, Sergeant Patrick Gass wrote that "one of the hunters killed two vultures, the largest fowls I had ever seen. I never saw any such as these except on the Columbia river and the seacoast."

On the return trip upriver, the party stopped on Deer Island, downstream from modern Portland. While some of the men caulked the canoes, others went to retrieve seven deer that advance hunters had shot earlier that morning. Reuben Fields soon returned with word that vultures and eagles had devoured four of the deer; the vultures had dragged one buck about thirty yards from the cache, skinned it, and broken its back. Joseph Fields shot one more big buzzard on April 6 near Multnomah Falls, and as the Corps continued east on the long journey home, their baggage included a condor skull and one long primary feather. The package arrived at Charles Willson Peale's museum in Philadelphia later that year, providing American naturalists with their first fragments of a bird that had been exciting the interest of visitors to the Pacific coast for two centuries.

Magnates of the Air ~

IN 1602 A CARMELITE FRIAR, standing on the deck of a Spanish ship, beheld orange-headed, black and white vultures circling a dead whale on a southern California beach. On shore in Monterey Bay, he saw huge birds "in the shape of turkeys" and estimated their wingspan at seventeen spans, or over eleven feet—a bit large even for a condor, but the friar was far from the last person to exaggerate their size. Over the next century, Franciscan missionaries journeying along the California coast dubbed the big birds "royal eagles" and witnessed their aggressive feeding habits and noisy struggles for position around a variety of reeking meals. The friars noticed that several California tribes raised the birds in captivity and used their feathers and stuffed bodies as effigies in ceremonies.

A royal eagle entered the cabinets of British science in 1792, when George Vancouver anchored in Monterey Bay. The ship's prickly naturalist, Archibald Menzies, promptly shot a huge buzzard and eventually delivered its skin to the British Museum, where it was examined, measured, and christened as a new species called the California vulture. A controversy erupted among ornithologists concerning the relationship between the Menzies specimen and the condor from the Andes of Peru and Chile, and the North American bird was ushered in and out of seven different genera until 1901, when it was finally designated *Gymnogyps californianus*, in reference to its naked head and home state.

The California condor, which seems to be such an unmistakably large and distinctly patterned bird, can be notoriously difficult to identify. They are often seen alone, flying at great heights. Their developing plumage, which takes half a dozen years to mature, includes stages of partially white underwing linings that can mimic both golden and bald eagles. As Thomas Jefferson understood, it often takes close examination to make a positive identification; that is why ornithologists are often reluctant to accept sight records alone as valid evidence. And yet, when trying to piece together the history of condors in the greater Northwest, there is room for the testimony of careful observers.

In the decades following Lewis and Clark's visit to the Columbia, sporadic mentions of large vultures continued to appear in the journals of fur traders and travelers. An agent on the Willamette River even complained of voracious buzzards dragging pine boughs off a cached deer and devouring the meat in a repeat of the Corps of Discovery's Deer Island experience. But none of the reports contained any description beyond size, and no one tracked the birds' movements or behavior until a succession of experienced naturalists arrived on the scene.

Traveling at the behest of the Horticultural Society of London, David Douglas landed at the mouth of the Columbia in the spring of 1825. Although trained as a botanist, Douglas applied his talents to birds and animals as well. This was an era when museums and scientific societies were dispatching collectors all over the world to acquire specimens, and Douglas immediately began sweeping up all manner of flora and fauna to send back to London. Large buzzards along the river soon caught his attention, but their extreme wariness made it hard to approach close enough to get a good shot. When he finally did succeed in bringing one down, his buckshot pierced its head, making it useless as a museum sample. Douglas entertained the idea of baiting one of the birds with a leg trap, but apparently never carried out this plan. He later shot both a male and a female near Multnomah Falls.

Douglas spent the winter of 1826–27 around Fort Vancouver, where he infected a clerk named George Barnston with a fervor for natural history. Barnston was especially impressed with the beautiful buzzards.

> This magnate of the air was ever hovoring around, wheeling in successive circles for a time, then changing the wing as if wishing to describe the figure 8; the end of the pinions, when near enough to be seen, having a bend waving upwards, all movements, whether floating or soaring, ascending or descending, are lines of beauty. In flight it is the most majestic bird I have ever seen.

Barnston sometimes startled the big vultures from well-rotted carcasses hidden in the forest, and he conjectured that the birds must use both sight and smell to direct them to carrion, and that in some cases the latter sense alone could suffice. Research on the buzzard family has since shown his deductions to be correct.

When a spell of unusually cold, snowy weather killed several horses around the trading post, Barnston watched with fascination as various scavengers descended on the dead stock. "Most conspicuous among these were the California vulture. . . . None may be seen in any direction, but in a few minutes after a horse or other large animal gives up the ghost they may be descried like specks in the aether, nearing by circles to their prey."

Douglas was also on hand, and his excitement reached a fever pitch when a hunter brought in a dead vulture. Barnston and his cohorts "had a hearty laugh at the eagerness with which the Botanist pounced upon it. In a very short time he had it almost in his embrace fathoming its stretch of wings. . . . This satisfied him, and the bird was carefully transferred to his studio for the purpose of being stuffed."

When Douglas departed the Columbia in March 1827, he carried two or three condor specimens and copious notes. He reckoned that their range extended as far north as the present Canadian border, but the birds were most abundant on the lower Columbia between the Dalles and the coast. He found enough in the Willamette Valley to class them as plentiful, including a group of nine birds that marked the largest single gathering he saw. "Great numbers" witnessed on the Umpqua River led Douglas to believe that the Columbia River vultures migrated south beyond the Willamette. "Preceding hurricanes or thunder-storms," he wrote, "they are seen most numerous and soar the highest."

In addition to these range delineations, the naturalist documented behavioral traits: "I have never seen them call except when fighting about food, when they jump trailing their wings on the ground, crying 'Crup Cra-a,' something like a common crow." As condors descend on a carcass, he wrote, "so gluttonous are they that they will eat until they can hardly walk and have been killed with a stick." After eating, they would often perch in a snag with their heads drawn deep into their ruffled collars and their wings hanging down over their feet.

Douglas reported that the Canadian voyageurs with whom he traveled on the river prized the long wing feathers for making pipe stems, although ornithologists have since questioned this assertion, noting that a finger-sized quill provides an unusually large diameter for a pipe tube. One of these voyageurs, Etienne Lucien, represented himself as an expert on the big vultures, claiming that they built stick nests in the thickest part of the forest and laid two jet-black eggs larger than those of a goose. Lucien was talking through his hat here: Condors do lay big eggs, but they shade more toward pale greens and pastel blues. The adults brood the eggs not in trees but on the rocky floors of caves and crevices. Douglas later published Lucien's account, injecting an element of fanciful lore into the slim annals of Columbia River condor history that proved difficult to erase.

━━ ≡◆≡ ━━

Next in the succession of visiting naturalists was John Kirk Townsend, who with his companion Thomas Nuttall accompanied Nathaniel Wyeth overland in 1834. From the moment they crossed the Continental Divide, Townsend was on buzzard alert: "I kept a sharp look-out for this rare and interesting bird in all situations on the route, which I thought likely to afford it a congenial dwelling place; but not one did I see."

In spring of 1835 or 1836, on a visit to the falls of the Willamette River, Townsend was watching a run of fish leaping at the rapids when he noticed dozens of turkey vultures sailing overhead. A rustling noise attracted his attention, "and there, to my inexpressible joy, soared the great Californian, seemingly intent upon watching the motions of his puny relatives below." The condor wheeled and plunged toward a freshly beached salmon. The naturalist fired, and the vulture fell on the opposite bank of the river.

An excited Townsend wasted no time in shedding his clothes and taking the plunge. A few vigorous strokes carried him across the current, and he sprang upon shore to secure his coveted specimen. But "the huge creature had been only wing-broken, and as I approached him, seemed determined not to yield himself a willing captive." Having left his gun with his clothes on the opposite shore, the naturalist looked around for a stick, but could not find one. In desperation he began pelting the condor with stones. Men,

49

women, children, and dogs who had been startled by the gunshot flocked from a nearby village to see what was going on.

For the next half hour, the naked man danced a stuttered tango with the injured bird, which "sometimes hobbled awkwardly away, when attacked, and at others dashed furiously at me, hissing like an angry serpent, and compelled me likewise to run." The laughter of the village women rang in Townsend's ears as he tried to kick sand in the attacking vulture's eyes. Finally the naturalist flung a lucky stone that plunked his quarry squarely in the head, and the stunned bird fell to the ground.

> In an instant I alighted upon him, sitting upon his body; and firmly grasping his neck with my hands. One of the Indians, at my request, brought me a knife, and I soon despatched him by severing the spine. I hired one of the boys to cross the river in a canoe to bring over my clothes and gun, and when dressed, skinned my prize with the Indians crowding around me, curious to see the operation.

At some point Townsend killed another condor near the mouth of the Columbia; one of these birds was a juvenile whose underwing feathers were not yet completely white. Upon his return to Philadelphia in November 1837, Townsend sold many of his specimens to Audubon, who extended the final volume of his *Birds of America* to include several "rara avis" brought home by the traveling ornithologist. Among the last dozen plates completed in the spring of 1838 was a painting of a single "Californian Vulture—Old Male," perched on a bare branch with its neck craning forward. One of the Columbia River condors almost certainly served as Audubon's model for a credible portrait of a bird he never saw in the wild.

Meanwhile, Townsend was being encouraged to write his own ornithology of the United States, intended to be less expensive and more accessible than Audubon's folios. In 1839 the young ornithologist produced a short facsimile including illustrations and descriptions of four vultures, but soon abandoned the project. However, his text for the California vulture appeared almost verbatim in the fifth volume of Audubon's *Ornithological Biography* with the acknowledgment that "Mr. Townsend, who has had opportunities of observing it, has favored me with the following account of its habits."

Townsend described the vulture's range as extending five hundred miles inland from the mouth of the Columbia, which would put it well into the Snake River territory. He said the birds appeared most frequently in the vicinity of cascades and falls, where they fed on dead salmon on the shore. He had often seen them near Indian villages, attracted by fish remains, and had watched a pair feeding on the carcass of a dead pig near Fort Vancouver. He had never seen their eggs, but Indians he had spoken with said the vultures bred in the alpine country and nested on the ground rather than in trees.

Townsend had watched them move on the ground in a "stately walk" and "a clumsy sort of hopping cantor . . . when about to rise, they always hop or run for several yards, in order to give an impetus to their heavy body." If he ever returned to the Columbia, Townsend concluded, he would try and capture a live vulture in a baited pen, after the method used in South America to snare Andean condors. But the ornithologist never returned to the West, and the two birds he carried back to Philadelphia became the last known condor specimens taken from the Northwest.

In 1845, Audubon gave a large portion of the skins used in his paintings to Spencer Baird, then a young professor of natural history and an ardent bird lover. When Baird installed his collection at the Smithsonian Institution a few years later, the vulture taken by John Townsend at the mouth of the Columbia River represented the museum's first specimen of a California condor.

Jesuit priest Modeste Demers arrived on the Columbia in 1838, and in succeeding years he traveled frequently between his station on the Cowlitz River and other missions in the region. In notes on the area's bird life penned in 1844, Demers listed a bulky black vulture that was said to be from California and was so well known for its sense of smell that the Cowlitz identified it as "who has a sharp nose." He repeated David Douglas's assertions that the odor of carrion attracted the buzzards from great distances, and that after gorging themselves they were so helpless they could be killed with a club. He also noted that "the feathers of this bird are much sought after by the aborigines, who after having tinted them in different colors attach them as ornaments to their long calumets."

If the vultures Father Demers saw along the Columbia during his treks in the early 1840s were indeed condors, he was witnessing a species on the decline. In the fall of 1841, scientists from the U.S. Exploring Expedition under Lieutenant Charles Wilkes covered the Columbia from its mouth far upstream without noting them any-where except the Willamette River. A decade later, career naturalists James Cooper and George Suckley combed the lower Columbia on a survey for the Northern Pacific Railroad, but only Cooper recorded one. In January 1854, during a relentless cold snap, he came upon a huge vulture perched high on a snag. As soon as the bird stretched out its long bare neck, Cooper knew he was looking at a condor. "The river being then frozen solid, and the ground covered with snow, it did not appear to admire the scenery and soon started off towards the south."

Cooper and Suckley regretted that they were not able to stay for fishing season to view the vultures at work, but it is far from cer-tain that time spent watching the great salmon runs at the Dalles or the Cascades would have rewarded them. After Cooper's single bird flew south, no naturalist ever reported seeing a condor on the lower Columbia again, in any season. In fact, there is only one other record generally accepted by ornithologists for the entire main stem of the river. That occurred far inland when C. Hart Merriam, a California biologist with plenty of condor experience, saw a single adult standing among the sagebrush east of Coulee City, Washington, in the fall of 1897.

Although condors were not recorded on the Columbia itself during the half century between Cooper's and Merriam's visits, sightings continued to dribble in from other parts of the Pacific Northwest. On the Canadian coast, they were spotted inside the city limits of Vancouver and at the mouth of the Fraser River. An 1893 bird list included a brief account of a pair landing on an island in the Fraser delta several years previous, but concluded with the words "none seen since, used to be common."

The Umpqua River drainage in west-central Oregon, where David Douglas had seen numerous condors, may have served as a last refuge for the birds in the region. When Titian Ramsey Peale traversed the Umpqua in 1841 as part of the Wilkes Expedition, he listed California vultures as part of the local scene along with ravens, crows, and Steller's jays. In an 1852 letter, Rosa Applegate Putnam, an Umpqua settler with an eye for all kinds of birds, described "the largest wild bird in the country . . . an overgrown buzzard—it only preys on the dead carcase—I saw one measured which I think was between ten & eleven feet from the point of one wing to the point of the other." Any vulture with a wingspan like that has to be a condor. In the summer of 1903, a father and son familiar with condors from California saw a pair of the great buzzards sailing aloft in the Umpqua Mountains. The following March they sighted another four flying much lower, almost within gunshot range, laboring against a stiff wind. That 1904 display drew the curtain on a bare century of historical data for *Gymnogyps* in the Northwest.

In their 1854 report on the natural history of the Northwest, Cooper and Suckley classified the condor as an irregular visitor to the region in spring and fall, and many ornithologists believed that the birds were wanderers who ventured out from their homeland in California. These could have taken off for the north, soaring high to catch the southwesterly winds that blew in with the big storms, sailing along the ridgetops of the coastal range. They would follow the contours of the beach north past the giant redwoods, pausing to feed on occasional carrion until they reached the plentiful offerings of a great river far away.

But perhaps these birds were not mere visitors. According to early chronicles, condors were present not only during spring and fall salmon runs, but all through the winter as well. Tribal sources told John Townsend in the 1830s that they had seen the birds' nests, but he recorded no details. Although no nesting site has ever been found in the Northwest, the fact that condors tend to stay close to home during their winter breeding season raises an intriguing possibility, and some scientists have conjectured that there were one or more small resident populations along the lower Columbia, the Willamette, and the Umpqua.

Whether the Northwest condors were residents or visitors, no one has a definitive explanation for why they disappeared from the Northwest so precipitously after European contact. Salmon continued to fight their way up the Columbia, and whales still washed up on Pacific beaches. Other sources of carrion, including new ones like horses and pigs, were readily available. It seems easy to blame the advent of buckshot, and even though the combined total of known specimens collected by visiting naturalists barely reached double digits, it is impossible to estimate how many birds were shot or trapped by curious furmen. Because condors do not breed until they are six years old, and then lay only one egg per year, a small increase in mortality could have tipped the balance for an isolated population. Whatever the ultimate cause, the beautiful buzzards recorded by those first fortunate visitors seem to have been caught in a narrowing spiral toward extinction.

Breakout ~

ON A SUNDAY MORNING in March 1971, David DeSante led a gaggle of fledgling bird-watchers toward a wooded lakeshore on the campus of Stanford University. As they approached the willows, he remarked on a flock of seventy-five or so turkey vultures cutting circles over the lake in slow pursuit of the day's first updraft. Among these buzzards, one bird immediately stood out, its wingspan stretching half again as wide as all the others.

DeSante told his group that they must be looking at an immature bald eagle, and the outsize raptor did indeed have some white under its wings. But it was so big. Puzzled, DeSante stopped talking and lifted his binoculars to his eyes. "Jesus," he whispered. "That's a condor."

As an ornithologist and devoted birder, DeSante knew exactly how unlikely an event this was; over the past thirty years, there had been only two other condor sightings in all of northern California, both near Monterey Bay. Aside from those anomalies, the dwindling population clung to the vicinity of the Sespe Condor Refuge, tucked into the dry coastal mountains of southern California. Only a few dozen of the birds remained, and a fair percentage of those had recently been bowled off their roost by a wicked winter storm. Since then the last survivors had circled their sanctuary as if trapped by the gravitational pull of a collapsing star.

Now, on a quiet Sunday morning, a condor had appeared over Palo Alto, more than three hundred miles north of the Sespe refuge. It floated within a gyre of turkey vultures, circling with such leisure that DeSante had time to disbelieve his eyes, shake with excitement, collect himself, then revert to a scientist's methodical process of identification. There was no doubt it was a condor. He made it to be a third-year bird, because the wing linings were beginning to fill out with white, and it had a good start on the collar of ruffled feathers that produces the classic ring-neck look. The head was drab, but DeSante knew that condors mature slowly, and that the citrus skin color might not assume its full glow for another couple of years.

After the buzzard flock gained altitude and began to disperse, the condor drifted off by itself in a northerly direction. Fifteen minutes later, as the news spread like wildfire among local bird enthusiasts, many small cars set out in dogged pursuit of the graceful soaring form. Beneath a thickening cloud cover, the chasers found and lost their prize several times. Toward evening a few lookouts were fairly certain they had seen a large bird circling down into a thick stand of cottonwood trees at the foot of a chaparral ridge.

Next morning, as a cold drizzle began to spread across the Bay Area, no one could locate the condor. Around lunchtime, DeSante coaxed some of his coworkers onto the roof of the Stanford biology building for an all-points search, but as the pace of the drizzle picked up, the watchers melted back into the comfort of the indoors. By 12:30 DeSante stood alone. It was then that the condor reappeared, gliding straight toward his head. The enormous form cruised past the rooftop almost at eye level, so close that he could count the tips of its outstretched primary feathers. He gulped hard, took a fix on its heading, and leaped back to the top floor to sound a second general alarm.

Only a few moments later, I happened to be climbing the steps toward DeSante's office when an agitated clump of humanity, my bird instructor at its center, swept me back down the stairway and into the parking lot. There, amidst much shouting of strategies and directions, frantic people piled into vehicles to pursue the errant bird. As they motored away, I tried to think about where a condor might hole up beneath the threat of a coming storm. I knew of a small lake nearby, nestled into a cow pasture that was strewn with radar dishes and blue oaks. It might, I decided, be worth a look.

The rain was spluttering down in a blustery rhythm by the time I arrived at the pond. A muskrat cut a smooth wake through the raindrops, and a raft of black coots waddled ashore to graze in the pasture. I scanned the power poles and radar towers for a dark shape, then made for a patch of open forest. This was a place where I had spent many hours searching for birds, and a tree-by-tree examination of the woods seemed completely reasonable. The wind died, and the raindrops began to pound.

In the closed gray world of oak shadows and crested wheatgrass, I paused to peel a piece of ragged bark from a madrone trunk, exposing smooth orange-red flesh stippled with knobs. Looking up into the torrent, I could picture an ancient mourner hunched among the branches of such a dripping tree. Its wings would be drooping low, and its naked head would be tucked into its ruff of neck feathers, lowered against the rain. But there was no condor in my madrone tree that afternoon. While I spent several hours engaged in a fruitless search, the rain gave way to a chilling wind. By the next morning the ridges around town were dusted with snow, lending a surreal air to the entire scene.

Soon after dawn, a determined pursuer located DeSante's condor just below the snowline, roosting glumly in a stand of eucalyptus trees. There it would sit until sunshine arrived to dry its soggy wings and stir the air upward. A line of watchers gathered along the roadside closest to the grove, keeping a respectful distance. Someone directed a portable telescope at the roosting tree. One by one the supplicants stooped to the eyepiece to watch raindrops trickle down the condor's pate, run the length of its upper beak, then follow the sharp downward curl of the hook to linger and fall off the tip. One of them opined that perhaps this vagrant had come to inspect a bit of its former territory for what could be the final time in the life span of the species.

When the sun broke through in late morning, the bird spread its great wings against the welcome rays. As each flight feather began to glisten black, the line along the roadside swelled. Smiling ushers pointed out the bird to people who never dreamed they would see a condor. At 11 A.M., the apparition ascended from the eucalyptus branch to take its leave. Last reports had it spiraling east, inland, making for the distant hills.

Lightning Birds~

THE PEAK OF VULTURE existence in North America coincided with the last glacial advance of the Pleistocene, roughly a hundred thousand years ago, when seven species of big buzzards, including *Gymnogyps californianus*, circled the skies. Fossil bones of California condors have emerged from spruce-pine woodlands in western New York state, central Florida's palmetto plains, Nevada's Great Basin sagebrush, and the cliff faces of Texas's Big Bend. Tissue clipped from a mummified condor skull found in a cave above Colorado's Grand Canyon has been dated back twelve thousand years.

The mammals of that epoch—ground sloths and mammoths, wandering herds of bison and camels, odd goats, three-toed horses, and deer with extravagant antlers—provided an abundant food supply. At water holes and kill sites, the assorted Ice Age vultures must have formed carcass mobs comparable to current scenes in Africa, where a variety of carrion eaters often operate side by side to clean up cat kills and drought mortality. But after the ice receded, a warming climate and changes in the vegetative mix, possibly aided by human hunters, extinguished many members of that grand mammal parade, and their disappearance sounded a death knell for all but one genus of the giant vultures. Somehow *Gymnogyps* hung on in western North America, a survivor from a different time.

In the mid-1950s, University of Oregon archaeologists, working at the head of the Dalles rapids on the Columbia, chiseled through a succession of sediment layers that showed signs of

58

steady human habitation. The older layers (carbon-dated to over eight thousand years before the present) yielded literally thousands of bird bones. When separated and arranged by species, the remains of more than 250 seagulls gave rise to the theory that people at the campsite had netted gulls for food. Bald eagles, which typically scavenge on dead fish, represented the next most common bird, followed by double-crested cormorants, strong-legged swimmers that still thrive along the river. The fourth most numerous species proved to be the California condor, with bones from at least sixty-three individuals identified. Remains of other scavenger birds included a dozen each of ravens and turkey vultures, three magpies, and a single wing bone from an extinct western vulture. The archaeologist who worked on the birds noted that many of the bones had been fractured before they were deposited, and he speculated about curious marks that appeared on some of them. After that initial analysis, the avian remains were placed in storage.

Half a century later, graduate student Victoria Hansel-Kuehn was leafing through a book collection left behind by her great-grandfather, an amateur archaeologist, when she found a report about the bird bones at the Dalles. After some preliminary investigations, she decided to reanalyze those remains for her master's thesis at Washington State University. In her work, she made a connection that her ancestor surely would have loved.

Victoria spreads the bones across a table in her lab in Pullman for me to admire. Time has mellowed their color to a warm, syrupy brown. Because they were preserved by a covering of exceptionally fine, dry, wind-blown sands, the bones remain surprisingly light and strong. Many of the shapes display an elegant symmetry—the cup of a gull shoulder socket; eagle eye rings in perfect small circles; the Slinky toy of a goose's windpipe. The outer surface of a condor's upper bill is embossed with a circuitous maze of grooves; its shape flows from the twin oblong portholes of the nostrils along a formidable ridge before turning ninety degrees to form an abattoir's hook. A condor's upper wing bone spreads as wide as a garden trowel and looks as though it could still carry the weight of flapping wings.

Because of the optimum condition of their preservation, each scratch on these bones remains sharply defined. Victoria holds up the short nub of a condor radius to show me a ragged, random break on one end, then a neat ring cut on the other. She picks up a humerus and points to an area near the joint marked with a series of parallel strikes that hold the same dark color as the surface of the bone. To her trained eye, both of these marks appear to have been made by stone tools.

During the course of her thesis project, Victoria examined over nine thousand bone fragments, and catalogued over three hundred marks that she judged to be the work of human tools. She consulted with other archaeologists, who confirmed her findings. Their general conclusion was that the gull and cormorant remains showed extensive butchery, indicating that these birds were probably utilized as food. Condor and eagle wing bones, on the other hand, showed a systematic pattern of tool marks that appeared to have been aimed at severing primary and secondary feathers. Indications that talons had been removed from several bald eagle metatarsals reinforced the notion that condors and eagles were being harvested for cultural purposes. And while three hundred out of nine thousand might not seem like a very high percentage, the fact is that very few culturally modified bird bones have ever been documented in the Northwest. Those parallel strikes on the wing bones laid out on the table have something to say about big vultures and people at the Dalles long, long ago.

— ·— ▆◆▆ —· —

The archaeological record from the Dalles is succeeded by linguistic evidence from many Northwest tribes. On the back cover of the field notebook that William Clark used during his winter at Fort

60

Clatsop, he copied his approximation of the Chinook word for buzzard, *E-pe-a*. The captain did not designate whether this word referred to his beautiful buzzard, but other tribes throughout the region had specific terms for big vultures or big eagles, and stories to go with them. Among several legends related to a visiting ethnologist in the 1890s by a Chinook elder was the tale of a young man who wanted wings in order to fly. He killed a succession of birds of progressively larger size, beginning with a young eagle, whose skin barely reached his knees. He then killed a white-headed eagle, and its skin reached a little lower, but it still was not large enough. Finally he shot a large bald-headed eagle, whose feathers provided a cloak that was just the right size. The "white-headed eagle" must have been the bird we now call the bald eagle, whereas the larger "bald-headed eagle" would logically have been the only bald-headed bird known to be larger than an eagle—the condor.

Wasco people, living on the Oregon side of the Columbia Gorge, refer to big birds that lived on the river long ago as "lightning birds." Some of their woven baskets are still decorated with a traditional motif of large birds with triangular wings outstretched; sometimes the birds' dark underwings are struck through with white lightning designs. North of the Columbia, an elder of the Snoqualmie tribe, which occupied the west slope of the Cascades, gave a turn-of-the-century anthropologist a word for condor that translated as "breaks down the weirs." On the east side of the Cascades, the Sahaptin-speaking Yakama people used a term meaning "crooked beak" or "rough-looking bird" to identify an enormous dark bird. The Cayuse, Walla Walla, and Nez Perce peoples farther upstream had distinct words to distinguish between small and large vultures. Although these tribes lived outside the area around the lower Columbia usually designated as condor range, their homelands contained plenty of spawning salmon and game animals to attract the sharp eyes and nose of a scavenger.

Cow on the Ground~

IN THE 1930S, anthropologist Claude Schaeffer, working among the Blackfeet tribes in Montana and Alberta, began compiling a list of tribal names for local bird species. After recording words for turkey vulture, golden eagle, and bald eagle, several of his informants volunteered yet another term, *omaxsapìtau*, which they said meant "big dark eagle." Curious, Schaeffer dug into bird journals and found an 1896 sighting of two condors in the Bow River Valley of southern Alberta, attributed to an associate at the Royal British Columbia Museum named John Fannin—the same man who had reported one of the last condors in the Fraser River Delta. "I was not aware that this bird was found east of the Rocky Mountains, or so far north," Fannin noted.

Using this sighting as a touchstone, Schaeffer began to ask questions about big birds along the East Front of the Rockies—exactly where a condor might spill out if it flew up the Columbia and caught an uplift over the Continental Divide, or rode the high winds across the Snake River Plain and turned north. Schaeffer found a surprising number of oral traditions relating to the omaxsapìtau.

A man from Browning, Montana, told about a Piegan named Big Crow, who was crossing a coulee with his wife when they spotted a large object up ahead that they thought at first was a cow. They soon realized they were looking at an enormous dark bird with a feathered ruff and a bald head. The bird launched itself into the air and flew off toward the Rocky Mountains. Big Crow had never seen anything like it before, and even though several of his friends did not believe him, the year was entered into the Blackfeet winter count as "that in which Big Crow saw the omaxsapìtau." Schaeffer and his informant calculated that this incident had occurred around 1897—the same year that C. H. Merriam saw a condor standing on the ground near Coulee City, and only a year after John Fannin's sighting north of the border in Alberta.

An eighty-seven-year-old man named Rides at the Door remembered seeing a "big eagle" while raiding for horses to the south. Chewing Back Bones said that sometime in the 1860s his father had called off a raid on the Crow Indians in southern Montana when an omaxsapìtau flew directly at his party. The parents of Dog Takes a Gun had told him about a big bird near Calgary that dropped a tail feather measuring two feet in length, and he himself knew of a curiosity dealer in Calgary who had kept an entire wing from a similar-sized bird for several years.

Harry Under Mouse told a story he had heard from his grandfather White Bear, who had been born around 1820. White Bear used to capture eagles for ceremonial purposes in the traditional Plains manner, by lying in a shallow pit covered with branches and clutching a chunk of meat to lure them in. When an eagle struck the bait, the hunter would grab the bird's legs and, dodging the razor-sharp beak, try to dash it to death on the ground. On one occasion, as he lay on his back in a pit, White Bear saw an immense bird circling over his head. It hovered warily, then landed on the ground nearby and hesitated for some time before finally approaching the bait. White Bear lay still until the bird came so close that he could see its hooked beak and coarse scales on its unfeathered legs. Fearing what would happen if he grabbed those legs, White Bear picked up a stick and frightened the bird away.

Yellow Kidney described a bird of extraordinary size, dark in color, with white under its wings and an elongated tail. Its large, hooked beak was a dark blue color that shaded to yellow at its base. He said such birds appeared infrequently in summer as far north as Montana and Alberta, but their home range lay somewhere to the south. There had been times when the omaxsapìtau had been spotted soaring high over Blackfeet camps in the foothills, circling in wide gyres before suddenly breaking off to sail westward, toward the spine of the Rockies. Piegan elders had told

Yellow Kidney that in former times the birds had been attracted to the leftovers of buffalo hunts. He had also heard that in those days, Blackfeet eagle hunters occasionally succeeded in capturing an omaxsapìtau. Then the ritual practices associated with a golden eagle would be transferred to the larger bird. All or part of its flesh and skin would be offered to the sun, and seven of the primary feathers would be attached in a semicircle, quills forming a point, on a specially prepared buffalo calf hide.

Bands of the Cree nation also hunted the northern Plains along the east slope of the Rockies, and Harry Under Mouse told Schaeffer that a group of Cree at the Hobbema Reserve south of Edmonton had kept a stuffed omaxsapìtau as regalia for years. In certain ceremonies they would tie the effigy to a performer's back and attach its wings to his arms so he could make the bird appear to fly. Its head loomed over the dancer's head, and the tail hung so low that it almost touched the ground.

Claude Schaeffer was not surprised when some of the lore he collected wandered into the realm of myth. Elders recalled two different dreamers, both appropriately named Big Eagle, who claimed omaxsapìtau as their spirit bird. Yellow Kidney said that on clear days the first Big Eagle could wave his ceremonial feathers at the sun and make a sun dog appear. Harry Under Mouse related that the second Big Eagle would wave his omaxsapìtau tail feather in a certain way to put adversaries to sleep. Once, when he and some companions were surrounded by angry Cree on the top of Devil's Head Mountain near Edmonton, this Big Eagle used the feather to render his entire party invisible so they could make their escape.

As Schaeffer analyzed the Blackfeet accounts of the omaxsapìtau, he paid careful attention to the possible confusion with eagles and turkey vultures, and to the exaggerations and inconsistencies that accumulate around any body of lore. He knew the Blackfeet to be sophisticated observers of all types of wildlife, and he concluded that the combination of traits they described—the naked head, neck ruffles, long wings, and unfeathered legs; the wary vulturelike approach to prey; the long spiral ascensions toward the sun—made a very convincing match with the California condor. He felt there was every reason to believe that tribesmen on the northern Plains "had long ago observed its great, sweeping flights from out of the fastness of the Rockies."

Specks in the Ether~

THE BIRD STANDS motionless for a long time, steps a bit to the left, and stands again. Its toes are big and reptilian, ending in blunt nails that are not at all like a raptor's talons. As its head turns this way and that, I can see that the dull orange skin of its crown is marbled with shades of yellow and red, and its wrinkled neck is marked with carbuncles. A surprisingly short bluish beak emerges from a fleshy sheath, then lightens to yellow at the tip. After a few minutes the bird hops to a dead branch, raising its great dark wings to reveal the striking white triangles underneath. The long primaries settle somewhat awkwardly as the condor returns to its stationary pose, but its neck remains unbowed. Only the large square tags clipped to the bend of its wings make it difficult to imagine this bird standing beside the glistening body of a spawning salmon, or surveying the torn flesh below a buffalo jump.

But this is a captive bird, afloat in an elusive space between extinct and wild. I am viewing it not across a foaming rapid but through a narrow window that forces me to bend and lurch until I achieve the same oddly hunched posture as the vulture. The high-ceilinged room is gloomily dark, designed to mimic the dank niches in which generations of condors have passed their time. A calf carcass lies on the floor, leaking the authentic smell of carrion. The condor's neck twists, then collapses into its charcoal ruff; the great naked head resumes its unmoving stance. Like other birds in similar dark rooms, this condor was reared on the outskirts of Boise,

Idaho, at the World Center for Birds of Prey, which conducts studies and conservation programs for raptor species in peril around the
globe. Because of their success with the captive breeding and
release of peregrine falcons, the center was asked to help with condor restoration efforts beginning in 1993.

Ten years earlier, the dwindling condor population at the
Sespe Refuge in southern California had reached a low of twenty-
two birds, and a controversy raged over whether it would be better
to intervene or to let the birds die in peace. After bitter arguments
from both positions, a decision was made to capture the handful of
birds remaining in the wild. Some people were ready to write the
condor's obituary the moment those last birds were wrestled into
nets; they deemed resurrecting an entire species from such low
numbers an impossible task.

It is a ticklish process to coax captive condors to mate and lay
eggs, to raise young in such a way that they will be able to survive
on their own. But recovery programs at the San Diego Wild
Animal Park and the Los Angeles Zoo have solved many aspects of
the puzzle, and over the last three decades, the number of living
condors has increased tenfold. The birds entrusted to the World
Center live in special housing designed to simulate the breeding
caves of wild condors, and remote video feeds allow keepers to
monitor behavior while minimizing contact with humans. One
thing recovery workers have learned is that birds being raised for
the wild gain nothing from being fussed over.

The Snake River Plain is a vast stretch of sagebrush that sweeps
from the Boise Mountains across the heart of Idaho into the
Owyhee Range. The river traces a wide arc, invisible until you
touch the rim of its awesome canyon and look down on cormorants
and white pelicans—so much water in a dry world. Most of the
plateau is public land, divided between grazing allotments and
artillery ranges. Large checkerboards of space have been set aside
for the Snake River Birds of Prey Area, where a walker on a cool

spring morning is much more likely to be buzzed by a prairie fal-
con than by a practicing jet. In the long silences between aerial
events, wandering feet continually kick up surprises—a delicate
lily blooming here, a rubber boa snake compacted into a protective
half-hitch there.

The World Center buildings on the edge of the refuge house
several dozen California condors. It seems fitting that the hope for
the future of the species lies here in the Boise Basin, for this was
the location of one of the last sightings for the entire Columbia
drainage. Reports of vultures of extraordinary size had filtered in
from the earliest days of exploration in the Snake River country,
and both David Douglas and John Townsend reported that condors
could be found in the mountains above the Snake, but no detailed
eyewitness account emerged from the area until 1918, when
General T. E. Wilcox submitted an oral statement to the
Biological Society of Washington, D.C., concerning an incident
from his time in Idaho Territory. In the fall of 1879, the general
said, near the hot springs above Boise City, he had happened upon
what he was certain were

> two California vultures . . . feeding on the carcass of a sheep.
> They hissed at me and ran along the ground for some distance
> before they were able to rise in flight. They were much larger
> than turkey buzzards, with which I was quite familiar, and I was
> very close to them so that I could not be mistaken in their iden-
> tity. The cattle-men said that the California vulture or buzzard
> was not uncommon there before they began to poison carcasses
> to kill wolves.

••• ≡◆≡ •••

Since 1996, several dozen condors raised in captivity have been
released in southern California and Arizona, and more are added
yearly in the hope of establishing self-sustaining wild populations.

Now the question is whether these ancient birds, like a very few species before them, can effectively reclaim lost territory. Perhaps it should be taken as a hopeful sign that some newly freed condors have shredded tents and sleeping bags in northern Arizona fishing camps, reminiscent of those feisty marauders that raided deer caches and broke through fishing weirs in the Northwest two hundred years ago.

If condors continue to reclaim their independence, it should be only a matter of time before a young bird takes off from the cliffs of southern California and heads north. The explorer could soar up high to catch the thermals and sail off along the ridge tops toward a place where the ocean tides fight the current of a big river into eddies and rips. With a pressure system at its back, such a bird could make the mouth of the Columbia River in a few days—what is nine hundred miles, after all, under the breadth of ten-foot wings? Then it could head upstream, riding the afternoon wind, scanning the banks for salmon bodies up through the Gorge, following the river's bends and tributaries across the arid lands to the Rocky Mountains and beyond.

But the legend must return, like Osiris, as potent as ever—not like milling hatchery salmon or ranch-fed beefalo. We want the noise-making birds to return, hissing and wheezing. We want them to tear at the sour-smelling blubber of a whale on the beach, to cruise rivers where fish flail against rocks, and to pound over the crest of the Rockies carrying lightning beneath their wings. We want to look up and wait expectantly for those hallowed specks in the ether to appear once more, descending in wide circles to the earth.

CHAPTER FIVE

Behemoth

Skull of Columbian mammoth *(Mammuthus columbi)*

A Peculiar Piece of Ground ~

A READER OF THE JUNE 10, 1876, edition of the *Walla Walla Union* would have learned that the Columbia River was still rising with spring runoff. A crew had been dispatched to repair the wagon road to Wallula, and plans were well underway for a Grand Centennial Fourth of July celebration. On page three, there was a report on rosebushes being despoiled in the local graveyard, coupled with a message to anyone with desecration in mind: "The ghoul in human shape that thus despoiled a grave is unfit to live, and at death should be buried in the center of the most public cross-road in this country." A warning that the cemetery was being watched was followed by a news item from a small settlement a hundred miles northeast on Hangman Creek, so called after a summary execution beside its banks during a conflict between local tribes and U.S. soldiers in 1858. The creek bisected a lush valley known by the traditional Coeur d'Alene name of Latah, which was tucked in the rolling grasslands of the Palouse Hills on the eastern edge of Washington Territory:

> LARGE HORN—A private letter from Hangman Creek, Stevens county, dated the 30th of May, states that parties living there have found some of the bones of a very large animal. . . . "We have just got out a piece of the horn 3 feet 7 1/2 inches long, being just a part of the point of the horn. The horn when we get it all out will measure 9 or 10 feet long."

Looking back, Alonzo Coplen thought it was simple curiosity that precipitated the great adventure that befell his family in the spring of 1876. The Coplen homestead was perched on a hillside overlooking a wide bend in Hangman Creek. Several springs clustered in the creek bottom, forming a boggy area where cattle occasionally got mired. Alonzo described a peculiar piece of ground raised a little above the level of the flat, a peat formation that shook when anyone walked across it.

One afternoon in May, the brothers began probing the ground near one of the springs with a long pole, and indeed, there is nothing quite like a bog to invite a good poke: the ooze; the trembling earth; the shimmers of anticipation at what might lie beneath. Their rod hissed down through the green mat of sedges. On one of the thrusts, it struck something hard. Their interest piqued, the brothers mucked back to the barnyard and attached a large iron hook to the end of their pole. They plunged this implement into the morass, grappled about, and after a time coaxed a large object to the surface. Upon examination it turned out to be an enormous vertebra, far too large for any animal they had ever seen. Back down went the grappling hook, and this time an outsize shoulder blade emerged, two feet long and almost as wide.

There was no stopping them now. Thirty-three-year-old Ben, who had worked a stint in the Colorado silver mines, outlined a plan to drain a section of the bog. Luckily, he had a crew of siblings to help: Lewis, who was twenty-nine; George, twenty; Alonzo, thirteen; and eight-year-old Isaac. Beginning on the shore of the creek, a hundred yards from their find, they set to work slicing a deep channel through the wetland. To keep the walls of the fresh ditch from collapsing, they drove stakes along both sides, then stuffed brush tight behind them. After digging through the black topsoil, they shoveled a layer of fibrous peat, then a stratum of white volcanic ash, then another peat layer that was speckled with woody debris.

At a depth of eight or ten feet, they bottomed out on a bed of gravel and began trenching steadfastly toward the targeted spring-hole. Alonzo recollected that about halfway there, they uncovered a large stone spearhead, several stone arrowheads, and a small human skull. Just beyond these startling finds, their shovels struck an area where the sand and gravel were blackened by what he and his brothers took to be ashes from a prehistoric campfire. The budding archaeologists collected the artifacts and kept digging, holding their course for the alluring spring.

Although the Latah valley was still only sparsely settled, a few neighbors pitched in to help as the excavation continued through the long days of June. When Ben's ambitious trench finally reached the vicinity of the original discoveries, the diggers hit a mother lode of curious bones. Some lay at the bottom of the seeping spring in a layer of blue clay, and were hard and firm; others were embedded in the surrounding peat. These looked perfect when first uncovered, but Ben characterized them as "soft like soap," and many crumbled to pieces when moved.

Before long, over a hundred bones lay drying on the ground, and it was soon apparent that more than one type of animal was buried in the bog. Ben thought some of the shapes had come from squirrels, while Alonzo remembered horselike parts and the skull of a bird with a long beak. To farm boys who had been around livestock carcasses all their lives, many of the larger bones were recognizable mammal parts like vertebrae and ribs. Not so familiar were several long, elegantly curved appendages, which the Coplens interpreted as horns, several orders of magnitude larger than those of an ox or long-horned cow.

The Coplens and their neighbors were not the first people to be puzzled by such "horns." Medieval Europeans who dug up similar objects believed that they must have adorned the heads of dragons or griffins. Mongol tribesmen in Siberia attributed the curved ivory they found along their riverbanks to enormous mud-eating rats called mammuts that lived underground and used the long horns to dig tunnels through the earth. Some scientists assigned them to antediluvian creatures that had missed the Ark for one reason or another. When seventeenth-century quarry workers dug up several of the strange appendages in a small town in Germany, the mayor declared that they had come from unicorns and even drew a

picture of what the beasts must have looked like. Travelers returning from India and Africa called attention to the similarity between
these long horns and the tusks of elephants, spawning explanations
for their presence in the north that ranged from animals left by
Hannibal's invasion to skeletons swept north by the Deluge.

As word of the discovery on Hangman Creek spread, settlers
arrived from miles around to gawk at the outrageous specimens; at
some point the brothers, feeding on the shared excitement,
decided they should exhibit this wonderful menagerie around the
countryside. While it wasn't out of character for any Coplen to
ramble, Ben in particular had displayed a pair of itchy feet. After
homesteading around the Midwest with his family, he had journeyed to Colorado at age eighteen for a go at hard-rock mining.
Just before the Civil War, he returned to Iowa and married, then
brought his bride back to Colorado. Over the next two years, Ben
gained a daughter, rode with the U.S. Cavalry, and buried his wife.
He sampled frontier life in Wyoming and Nevada before rejoining
his father's clan, who by that time was settling into a third homestead in Washington Territory. A burly man with wild eyebrows,
Ben was by all accounts an engaging and gregarious character; he
was probably the instigator for the tour. Brother Lewis pitched in to
help, and the entrepreneurs loaded a farm wagon with the best of
their big bones and tossed a tent on top.

Grand Discoveries ~

THE BROTHERS RATTLED west and south thirty miles through the heart
of the Palouse to the small town of Colfax. Although it was only a
half-dozen years old, its population already numbered more than
a hundred souls, and the town boasted a flour mill, a sawmill, two
general stores, and two kerosene streetlights. Apparently one of the
first things Ben and Lewis did when they arrived on June 26 was
find a scale and enlist volunteers to help them weigh and measure

their bones. The "horn" they had brought weighed 145 pounds and measured 10 feet along the outside of its curve. The pelvis tipped the scales at 135 pounds, followed by a jawbone, 63 pounds, and a shoulder blade, 40 pounds. A single small tooth, only half the size of the ones still set in the jaw, weighed 10 pounds.

Some of the onlookers in Colfax that weekend were sufficiently inspired by the exhibit to fire off letters to their favorite newspapers. J. H. Kenedy, who helped weigh and measure the collection, pronounced it "the grandest discovery of the age to the geological world" and postulated that the Coplens had unearthed "an animal known to the antiquists as the behemoth." The author of an unsigned letter in the *Walla Walla Union* consulted a more modern text, using the adjective "mammoth" to describe the bones. The pelvis was so enormous, he marveled, that a grown man could pass through its opening by stooping just a little. He correctly noted that the flat grinding surfaces of the teeth differed from those of mastodons, which have pronounced knobs, like the bottom of an egg carton, on their biting surfaces.

Local schoolteacher James Edmiston also showed some knowledge of extinct pachyderms. He wondered whether the curved "horn" might actually be a tusk, and he noticed that the tip of the massive jawbone ended in a protuberance that tilted downward, as if to support a snout. His letter to the *Portland Oregonian*, headlined "The Centennial Mammoth," made a symbolic connection between the emergence of the magnificent bones and the celebration of America's first great anniversary. Using scientific terms like "processes" and "cartilaginous surface," Edmiston delineated the perfect half-moon opening of the pelvic girdle, but his letter was not all technical jargon. After recounting the massive thickness of one mandible, he indulged in a little wordplay: "This may be more jawbone than you like to take, but existing facts cannot be avoided."

With summer's dust swirling around their wagon, Ben and

Lewis continued south, ferried across the swollen Snake River, and made for the town of Dayton, where an amateur photographer took pictures of the bones. Like the Colfax correspondents, this shutter-bug had a keen interest in scientific matters, and as soon as he developed his plates, he dispatched copies east via the overland stage, addressed to Professor James Dwight Dana, the preeminent geologist at Yale University. Dana's *Manual of Geology*, a textbook used by basic science classes in colleges around the United States, included a section on extinct elephants.

The Coplens arrived in Walla Walla just in time for the largest Fourth of July centennial celebration in the Inland Northwest, with a crowd estimated at three to five thousand people. After a brass band and glee club performance of patriotic odes, a parade marched down Main Street, led by a small monkey turning hand-springs, marshaled by a man in a gorilla suit, and gravely followed by a baby elephant. Somewhere amidst the fanfare, another pho-tographer discovered the Coplens' own elephantine display. Within a few days, fresh prints labeled "the biggest 'horn' ever taken" and "the biggest jawbone in the country" graced the front window of his gallery.

While news of General Custer's defeat at the Little Bighorn began to rumble through the countryside, Ben and Lewis made the long wagon journey back to Hangman Creek and began work on a new pit a short distance from their original dig. A few days later, James Edmiston arrived to write a promised follow-up to his Colfax letter. There were new fossils to admire, including a rib almost five feet in length and a thirteen-foot tusk that was too rot-ten to extract. The inquisitive Edmiston must have been doing some homework in the weeks since he had viewed the bones in Colfax, because he now concluded that the animal belonged to the species *Elephas primigenius* (first-born elephant). This was the name that the scientific community had assigned to woolly mammoths at that time; it would be many years before all mammoths assumed their current genus name of *Mammuthus* and the creatures found in the Palouse were designated as *Mammuthus columbi*, Columbian mammoths. In addition to the bones, Edmiston noted that "a piece of charcoal was found on the gravel bed, bearing marks of an instrument. The marks are very distinct and seem to have been done with some rude instrument."

Edmiston also examined additional teeth that had been unearthed and remarked on their resemblance to those of living Indian elephants, "particularly in that the ribands of the teeth are waving and running obliquely crosswise." Elephant molars have always provided an important key for distinguishing species, but their variations in shape and size can be quite confusing. Looking at two of the loose teeth, Edmiston tried to describe them for his readers: "I can think of nothing better to compare them to than the head of a sucker fish, the top of the head representing the root of the tooth, the mouth the surface that has been used for nipping only."

⋯ ⚌✦⚌ ⋯

When a friend recently loaned us a beautiful mammoth molar she had found, we installed it in the center of the dining room table to admire. My kids could think of nothing better to compare it to than one of their mother's tennis shoes—the bottom of the sole representing the surface for chewing and the size being just about the same—but visitors came up with much more exotic likenesses. A math teacher took a quick look and pronounced it to be a sea urchin, pure and simple. Her son thought it looked more like a piece of whale baleen. The five-year-old entomologist next door was certain it was some kind of funny bug. None of them, however, suggested that it might be the tooth of a giant, the most common conclusion for observers of similar teeth for many generations, from an Englishman on the Essex shore in 1400 to the governor of Massachusetts Colony in 1705. African-born slaves in the Carolinas, however, immediately recognized three large molars unearthed there in the 1730s as "the grinders of an elephant."

Like a modern elephant, a baby mammoth was born with four molars, one in each quadrant of its jaw. These baby teeth couldn't process enough food to keep a young mammoth growing for long, and before the youngster was weaned at two or three years of age, a larger set of grinders erupted behind them, instead of beneath them as in other mammals. As the animal grew, these new teeth began to move forward in the jaw like a package on a conveyor belt, eventually pushing the worn first molars out of the mouth. A few years later a third set of molars, almost six inches long, emerged behind the second. It was undoubtably one of these early molars that a Colfax observer described as "the tooth of a human being, which is larger than three or four of our teeth." During the first three decades of a mammoth's life, six succeeding sets of molars rode into position, each bigger than the preceding set. The last and largest quartet of teeth, often measuring an honest foot in length, arrived when a mammoth was in its mid-thirties.

Mammoth baby teeth were similar in size and shape to human wisdom teeth, but the later teeth were something else entirely. On the biting surface of one of these molars, a narrow ivory ridge loops back and forth across the oval surface, like a long enamel worm curling gracefully back on itself in a continuous, controlled pattern. Succeeding teeth display increasing numbers of loops, with complex variations of amplitude and frequency; if you view the teeth from the side, you can see that these loops are formed by parallel plates of enamel, with thin layers of dentine sandwiched between. The number of enamel ridges, combined with the size of the tooth, provides biologists with a rough index for aging fossil finds; the Coplens, for example, unearthed animals ranging from youngsters to middle-aged adults.

When the animal chewed, the top and bottom molars slid back and forth against each other like millstones, masticating tough grass between the enamel ridges. The mandibles were in almost constant motion as the mammoth processed the enormous amount of roughage needed to support its massive weight; one Columbian mammoth, weighing up to ten tons, may have spent twenty hours a day ingesting an estimated five hundred pounds of raw vegetation. A short list of food items recovered from fossilized mammoth dung includes a high percentage of grasses, rushes, and sedges, supplemented with browse such as birch, rose, saltbush, sagebrush, blue spruce, wolfberry, and red-osier dogwood. Such a diet, and the dirt that came with it, would have taken a toll on any set of teeth. At some point not too far off the human allotment of three score and ten years, the ridged surfaces of those final four grinders were worn to the root and could no longer sufficiently masticate food, and many elderly animals apparently died of starvation.

⋅+ ▰◆▰ +⋅

Around the same time that James Edmiston visited the Latah dig, Philip Ritz, a commissioner assigned to collect grain samples for the Centennial Exposition in Philadelphia, also stopped by for a look at "the great unknown bones." He found that the brothers' new pit was yielding fossils only four feet below the surface. A pair of tusks looked to be nearly twelve feet long, but only one could be extracted whole. This brought the tusk count to nine, ranging in length from three to twelve feet. The brothers had also uncovered a huge skull, but it was too rotten to move and was left in "the black, oozy mud where it was deposited ages ago." Before concluding his letter, Ritz remarked that some other large bones recently had been discovered nearby.

William and Thomas Donahoe, Irish immigrants from Quebec, lived on a homestead about twelve miles west of the Latah valley. Tom recalled that when he and his brother Bill first heard about the Coplen discovery, they immediately thought of their own spring beside Pine Creek. About twenty feet deep with a clean sandy bottom, it had provided them with water since their arrival in the Palouse five years earlier. The brothers bolted sturdy grapnels onto two long poles and began to probe, Coplen-style, and it wasn't long before they hooked onto something solid beneath the sand.

For two days the Donahoes tugged and pulled, constructing a farmer's arsenal of levers and gantries and calling on neighbors to help. They finally extracted a gargantuan skull that measured thirty-five inches between the eyes, fifty inches between the ears, and forty-two inches from the back of the head to the front of the nose. Even without its tusks, which had broken off during the pulling and tugging, the Donahoes estimated the weight of the skull at eight hundred pounds. With seven able-bodied neighbors, they tried to haul up the rest of the skeleton, but the task was too much for their grapnels. "We found that would not do, so we started to ditch it," the brothers reported in a letter to Walla Walla. "We expect to have it ditched and the bones out in 8 days."

So another set of brothers found themselves astonished and eager to see more. Another assortment of amazing bones piled up around a spring. Another set of curious onlookers gathered around. "About the time we had most of the bones fished out," Tom remembered, "here comes a fellow with a band of sheep, headed for Montana. When he saw those bones, he just went wild, mind you. He wanted to trade those sheep, 700 of them, for a third interest in the elephant bones. He had a partner and we said he'd better wait and talk it over with him before he made a deal, but he was dead set on getting in on the fortune from the fossils. He was lucky we didn't take him up."

It is hard to say who was the luckier party; the next week, a Walla Walla paper reported that the price of sheep had reached its lowest level in ten years. But if the Donahoes hesitated to jump at the first offer dangled in front of them, they were not immune to the idea of commerce. Tom told a visitor that they were confident they were sitting on a bonanza, and by the end of the summer a second set of brothers was thinking of mounting a tour.

Antediluvian Bones ~

WHILE THE DONAHOES were hoisting the giant skull from their spring, Ben Coplen rolled back into Walla Walla to show off the most recent diggings from the Latah bog, including ribs, vertebrae, and five new tusks. "Our country is famous for bones if not money," crowed a local editor as reports of the two incredible discoveries continued to flash through regional newspapers.

The Coplen display gained further acclaim when Yale professor James Dwight Dana acknowledged the photographs sent to him from Dayton. In his reply, which was printed in the *Walla Walla Union*, Dana bestowed the first validation from the scientific establishment, stating that the bones belonged to "the extinct American Elephant or Mammoth." But a different opinion was printed the same day in western Oregon by the *Eugene City Guard*, whose editor had been following reports of the Palouse fossil finds and had concluded that the prehistoric artifacts certainly belonged to a unicorn.

By mid-August, Ben Coplen, aiming for larger crowds, had arranged with the vice president of the Oregon Steam and Navigation Company to provide free passage on a steamboat down the Columbia. When Ben boarded the steamer at Wallula Landing, at least one other person was along to help heft the massive fossils—probably his younger brother George, who was to enroll that fall in a secondary school near Portland, and possibly a neighbor named Bill Bohard as well. (According to Alonzo, Bohard had purchased Lewis Coplen's share in the enterprise for a span of horses and a small sum of money.) Whoever they were, they billed themselves as the Coplen Brothers, and downstream at The Dalles they found an empty storefront on Main Street in which to display their wares. After a week of business at the busy portage for river traffic, the exhibitors placed a brief ad in a local paper:

The Coplen Brothers wishing to give the children of this City an
opportunity to see the great fossil bones, extend an invitation to
all the Sabbath School children and their teachers to visit them,
free of charge, on this Saturday afternoon.

Then it was back on a steamer for the ride downstream to
Portland, where they set up their exhibit downtown, charging spec-
tators twenty-five cents a head. A thoughtful newspaper review of
the display compared the fossil-laden ground of eastern
Washington with Siberia's frozen taiga, which had recently yielded
an entire woolly mammoth encased in ice; like many a later scribe,
the reporter pondered the roles of climatic change and early man
in the animals' extinction. Mammoths and climate were also on
the mind of physician Philip Harvey, who accepted the loan of several bones from the Coplen collection to illustrate a lecture at the
Good Templars' Hall on the "Great Ice Time."

The last week of September was county fair time, and the
Coplens journeyed a few miles west to Hillsboro, where they were
welcomed by a newspaper editor as presenters of a first-class exhibition well worth patronizing: "A half hour spent in contemplating
these monster evidences of the earth's life, will give any one a better idea of the world we live in." Their competition included such
attractions as a world-renowned fire-eater, an eight-hundred-pound
woman, and Montgomery Queen's Circus. There was a display
known as 100,000 Curiosities, under the command of T. A. Wood,
who ran a combination natural history museum and freak show in
downtown Portland. Dr. J. J. McBride, The King of Pain, was
reputedly getting rich peddling a patent medicine known as The
World's Relief. A state senator took in the sights and crawled
through the pelvis of the unearthed behemoth, but a reporter
lamented that the fossils were not drawing all the attention they
deserved: "Their exhibition is something that everyone can
remember as food for profitable thought, which is more than
can be said of the majority of shows."

From Hillsboro, the Coplens moved on to Salem and a most appreciative audience in the person of Thomas Condon. A former Presbyterian missionary who was known around the territory for his paleontological zeal, Condon had recently resigned his position as Oregon's first state geologist to chair the science department at the new university in Eugene City. The professor, who had discovered mammoth remains himself and had looked at other finds in the region, declared the Palouse specimens "unquestionably the finest elephant remains ever unearthed on this Coast." He was especially impressed with the examples of teeth and jawbones, calling attention to one mandible in particular, in which "the fifth molar is standing alone in marvelous perfection of its minutest outlines."

In pondering the circumstances under which so many animals could have perished in an area not much larger than twenty feet square, the professor recounted his own experience with a packhorse that had wandered into a small peat bog and almost drowned in the bottomless mire before being rescued. A mammoth, he extrapolated, could easily have suffered the same fate.

> An elephant might have gone to the brink of that pool to drink and dropped into it as the horse did, and remained there as helpless as he, till he drowned. Another and another might follow, and the bones of these animals, thus trapped, remain buried in the oozy mud, thoroughly preserved from decay.

Condon, an imaginative and poetic geologist, went on to detail his vision of the Northwest landscape back in the days "when these, now fossil bones, were clothed with flesh," a time he reckoned at about two hundred thousand years ago.

※✦※

A century and a quarter later, Condon's life history conclusions remain remarkably apt, although his dates have been refined.

Scientists now reckon that primitive proboscideans, which included ancestors of animals as various as hyrax and dugongs, first appeared in the tropical environs of Africa. Mastodons branched off more than twenty-five million years ago, while the first true elephants, with their ridged molar teeth and dentine tusks, arose about six million years ago. Over the next million years or so, the family split into the similar but clearly distinguishable lines of African elephants, Asian elephants, and mammoths. These early mammoths, which developed a distinctive spiral to their tusks, spread north into the more temperate climates of Europe and Eurasia.

When the Ice Ages began around a million and a half years ago, ancestral mammoths, like so much of our fauna, moved across Beringia to North America. In a classic example of parallel evolution, the animals that remained in Eurasia gave rise to woolly mammoths, while their North American counterparts evolved into the Columbian mammoth. For some reason, Columbian mammoths grew to be larger than their Eurasian cousins. Their tusks took on an extra curl that gave them their distinctive lyre shape, but their teeth retained a simpler form, and paleontologists do not believe that their coats ever approached the shagginess of their woolly kin. Columbian mammoths wandered across the grassy meadowlands and open woods known as the mammoth steppe, from southern Canada deep into Mexico. Woolly mammoths eventually crossed to North America as well, but their bones are generally found in more northerly latitudes, following the advances and recessions of the great ice sheets. In some areas, such as the upper Columbia drainage, the ranges of the two beasts overlapped. As the chill of the Pleistocene oscillated toward a warmer and drier time, the numbers of both mammoth species began to wane, and the animals disappeared from all but the most isolated parts of their range around eleven thousand years ago.

Ballyhoo ~

THE WALLA WALLA County Fair ran during the same week as its Hillsboro counterpart and numbered among its exhibits the fabulous fossils of the Donahoe brothers, who had rented a booth and hired a "ballyhooer" to bark in the crowds. The ponderous skull provided the main attraction, but William and Tom also had wired together two leg bones for full effect, leading one reporter to speculate that the live animal "must have been two sizes larger than the Court House." When the fair was over, Tom Donahoe figured they had made about a hundred and twenty dollars for their efforts, and neither he nor the newly married William was interested in pursuing the traveling life any further. The brothers therefore sold their fossils for seven hundred dollars—about the same price as a herd of sheep—to "traveling agents" Falkner, Mitchell, and Thwing.

The new owners announced plans for a winter tour through California, and Nathaniel Thwing, who seems to have been the primary promoter for the tour, packed the bones and booked steamboat passage to Portland. Later characterized by a newspaper editor as an adept showman, Thwing demonstrated a talent for hyperbole, informing an editor at The Dalles that the tusks in his collection measured twenty-five feet in length—about twice the span of any known mammoth tusk. He was equally creative in matters of taxonomy, identifying his titan as a "mammoth mastodon." (He did not specify whether this indicated an extra-large mastodon or a hybrid proboscidean.) By the time he reached Portland, the leg bones had grown to twenty feet in length, the weight of one tooth had more than doubled, and the skull, containing "gold and silver quartz in profusion," was touted as the largest fossil remain ever found. Thwing had also decided to make a detour before mounting his tour of California, and he made sure that everyone knew his bones were headed to the great Centennial Fair, which pundits predicted would attract the largest crowd ever congregated in Oregon.

When the Centennial Fair opened in Salem on October 10, 1876, familiar attractions such as Montgomery Queen's Centennial on Wheels and his Aggregation of Transcendental Elegance were on hand, along with T. A. Wood's 100,000 Curiosities. So were two displays of mammoth bones. The Coplens handled their collection personally, while Nathaniel Thwing and a Walla Walla associate named John Hancock shilled for the Donahoe specimens. The fossil extravaganzas gained numerous mentions in the Fairground Jottings of local papers; several referred to the bones' appeal to scientists and deep thinkers. One such cogitator was a Dr. Davis, identified as "the philosopher of Harrisburg," who confided to a reporter that "the bones of the pre-historic brute found in Washington Territory . . . came from the moon when the continent of Africa crashed down after a volcanic eruption."

As the fair came to a close, the "rival big bones men" headed off in opposite directions. Thwing traveled north to Portland, where the "mammoth mastodon" was loaded aboard a steamer to San Francisco. In early November the *San Francisco Daily Call* announced the opening of Thwing's exhibit, reputed to contain seven distinct species of an enormous animal resembling the mammoth; the editor lamented that "neither the exhibitor nor the reporter is equal to giving a lucid description of the fossils."

The Coplen Brothers, meanwhile, moved south on a tour of the Willamette Valley. By October 20, their display was installed in a former hat shop in downtown Albany and the *Eugene City Guard* was anticipating the imminent arrival of the Coplen bones in that town. Then something happened to change Ben's plans. Perhaps it was the webfoot weather—it had rained every day since the fair ended—but the November 4 edition of the *Guard* included a succinct and disappointing headline: "Not Coming." A couple of days later, it was announced that the great fossil bones had been leased to Pacific University in Forest Grove for use in its geology classes;

within two weeks of this notice, the Tualatin Academy, a prepara-
tory school adjacent to Pacific University, listed George Coplen of
Washington Territory among its fall enrollees. The following winter, Ben Coplen confided to a visitor that custody of the bones had
been exchanged for tuition.

Farmer's Luck ~

IN THE LATE FALL OF 1877, a young fossil collector named Charles H.
Sternberg was traveling down the Columbia River when he met an
army surgeon who told him about the discoveries of mammoth
bones in eastern Washington. Determined to investigate the fantas-
tic discoveries, Sternberg journeyed to Walla Walla, where he
hired two helpers and some packhorses and headed north. The
men worked their way up Pine Creek to a promising spring, where
they dug through layers of peat and clay before hitting a bed of
gravel. Weeks of tedious, wet work produced a number of fine buffalo skulls, but not a single mammoth bone. "The farmer-fossil-
hunters had been more fortunate," Sternberg wrote.

Looking for some of that farmer's luck, Sternberg visited both
the Coplen and Donahoe springs. He listened to the list of animal
remains taken from the digs, including deer, buffalo, and birds. He
noted with interest the spear point that the Coplens had found on
the gravel, and the tool marks on a piece of charred and partially
petrified wood. Pondering some flint arrowheads and a bone tool
that he had found mingled with the bison bones from his own dig,
Sternberg wrote:

> I never doubted, from what I saw and heard at the other excava-
> tions in the immediate neighborhood, and where the collectors
> went through the same kind of peat, clay and gravel as we had
> gone through, that man, the buffalo, elephant, and many exist-
> ing species once lived together in eastern Washington.

While Sternberg was correct in his assumptions about the coexistence of people and Pleistocene mammals in the Northwest, it is impossible to draw any definitive conclusions about the human artifacts discovered by the Coplens or Sternberg because of the unscientific nature of their excavations. Over the past century, distinctive fluted spearpoints identified as belonging to the Clovis period (between 11,500 and 11,000 years before the present) have been found near the remains of mammoths at more than a dozen sites. Very few of these finds show indisputable evidence of hunting, and the exact relationship between people and mammoths remains a subject of some controversy.

One group of theorists interprets the evidence as saying that mammoths, as well as other large mammals, were driven to extinction by human hunting. An alternate theory points to well-documented vegetative changes during the long warming trend that followed the ice's recession. In many areas, grasslands were replaced by thick forests, an impossible habitat for bulky mammoths. In the open country that did remain, a new vegetative mix may not have supplied the animals with the nutrition needed to sustain reproduction.

Most scientists today combine these two theories, painting a picture of diminishing populations of large mammals competing for water and food. Opportunistic hunting by humans, coupled with environmental stresses and a slow birth rate, may have pushed not only mammoths but other large grazing species such as camels and ground sloths past the point of recovery. A more complete understanding of the role of humans in mammoth extinction awaits further study; in the words of Thomas Condon, "A very lively interest will ever cluster around every new discovery of these fossils until this question is definitely settled."

When Ben Coplen left Portland in the fall of 1876, he announced his intention to return home and continue excavating in preparation for a tour back East. Neighbors of the Coplens later recalled helping Ben with additional digs for a couple of years after his original find, and Alonzo remembered that he and Isaac and neighboring children played with bones stored in the barn, but apparently the plan for the tour never came to fruition. In 1879, a graduate student at McGill University exhibited a molar from the Coplen collection at a meeting of the Natural History Society of Montreal and read a paper comparing the specimens from Washington Territory to similar bones found in Canada, but the rest of the colossal fossils remained at Pacific University, according to W. D. Lyman, a teacher there and a friend of George Coplen. In an article written after a visit to Hangman Creek that same year, Lyman noted that it was still "the intention of the discoverers to pursue the investigation, so that specimens even more extraordinary than those already found may reward their search." Whether there were no more spectacular bones to be found or whether other projects intervened is not known, but by 1883 Ben, Alonzo, and George were busy staking claims in the Coeur d'Alene silver district.

In the spring of 1886, the Chicago Academy of Sciences acquired seven hundred pounds of the Coplen mammoth bones for three hundred and fifty dollars. The cache included remains from four adults and the pelvis of a possibly unborn calf, but less than a whole animal. Over the course of the summer, Academy preparators sorted through the collection with the goal of a complete reconstruction. Leg bones were almost completely lacking; there were no lower teeth, and only a few fragments of a skull. The crew fashioned missing parts from plaster, using the remains of an Indian elephant as a model. Unveiled that fall, the skeleton represented the first full mount of a mammoth in North America.

At the Chicago Academy of Sciences' meeting in October 1886, Professor W. K. Higley listed the most impressive measurements of the reconstruction, which stood thirteen feet tall. "It will be of interest," the professor continued, "to compare the height of a few elephantine forms." The famous Siberian woolly mammoth mounted in St. Petersburg rose a mere nine feet three inches. P. T. Barnum's great attraction, the African elephant Jumbo, had been struck by a train in Toronto the previous year, and the circus magnate had assembled both a skeletal mount and a stuffed version to go back on the road. But even in his reconstructed glory, Jumbo measured only ten feet, three inches at the shoulder. A mounted Indian elephant at Northwestern University stood ten feet, eight inches. A giant mastodon displayed in Boston stretched to eleven feet. Therefore, Higley announced, the Chicago Academy of Sciences was the proud possessor of the largest known elephant in the world.

Around the same time that Higley was assembling his mount, a Spokane businessman named Isaac Peyton sought out Ben Coplen for an interview. Ben's answers were informed and thorough, beginning with a report on the heavy concentrations of iron and nickel in the fossil-bearing spring. He said he had probed several additional springs around his original site and had found that four or five of them contained bones; he had taken ribs out of at least one of them, confirming his belief that "the quantity of bones remaining unexcavated is very great." In a cryptic conclusion to notes that he forwarded to the president of the Chicago Academy, Peyton noted, "B. F. Coplen . . . Assist—probably charge nothing." For Ben, evidently, the discovery had never lost its appeal. Later that winter, he regaled W. M. Lee, a visitor from Tacoma, with details of the original dig, took him to view the bog, and presented him with several souvenirs. Ben said that they had taken out "remains of a cave bear, and hyenas, extinct birds, and a sea turtle" in addition to parts of nine mammoths. There were still enough tusks around the homestead to overwhelm Lee with their magnitude. Some of them were worn away several inches deep at the bottom of their curve, he noticed, as if they had been constantly rubbed on the ground. "Just imagine," Lee wrote, "far back in the misty bygones of antiquity, probably before the appearance of man on earth, that Washington territory was the home of the monstrous animals that roamed over the great prairies and traversed the Columbia River."

⚬⚬ ≋◆≋ ⚬⚬

There is no record of Benjamin Coplen or any of his brothers ever traveling to Chicago to admire the fruit of their labors. When the Latah mammoth served as the centerpiece of the Washington Pavilion at Chicago's World Columbian Exposition of 1893, Ben was celebrating the birth of a son and beginning his term as the first mayor of the newly incorporated town of Latah. He later tried the hotel business and, in 1901, moved with his family to Colorado Springs, Colorado. In December of that year, a Spokane newspaper ran an obituary lamenting that Latah's honored citizen had been run over by a train in Colorado. The article was in error; Ben bounced up from the grave and returned to Latah.

In 1907 he moved across the state line to Plummer, Idaho, where his two sons worked for the railroad. Alonzo Coplen was pretty sure Ben took only a few of the mammoth parts with him, and that the last of the great bones were left behind in the homestead barn. It was in Plummer that Ben passed away for keeps in 1912. Two years later, the newly formed Field Museum of Natural History in Chicago purchased the Latah mammoth skeleton from the Chicago Academy and had it boxed and stored. After a half dozen years in storage, it was remounted and placed in the front hall of the Field Museum's own grand exhibition.

While neither of the remarkable mammoth finds from the Palouse in the summer of 1876 fulfilled Thomas Condon's wish that they remain in the Northwest, both have lived up to the Colfax observers' prediction that they would prove to be of interest to science. The beautiful skull and other parts discovered by the Donahoe brothers on Pine Creek and carried to San Francisco by Nathaniel Thwing were purchased in 1878 by the paleontologist Edward Cope of Philadelphia. Two decades later, without ever opening the shipping crate, Cope sold them to the American Museum of Natural History in New York. In the 1920s and 1930s, taxonomists there measured and remeasured the Donahoe bones as they sorted out the different species of North American mammoths. Today the gigantic skull resides in the museum's basement as landmark specimen No. 8681.

The Latah mammoth, meanwhile, has remained in sight and on the move. In the 1950s, Field Museum preparators completely reconstructed the skeleton, curling its metatarsals to assume a more natural, springy posture; in 1993, the mount was sawn apart and carried to a new exhibit space, where it was assembled yet again. There it was recently studied by a paleophysician who measured calcification in the bones and determined that the animal had suffered from arthritis in its toes. During museum hours the Coplen mammoth still occupies a central position in the Hall of Time; a crowd of kids is often gathered around the reconstruction, staring up at its impossible size.

CHAPTER SIX

The Devouring
Disorder

Deserted Salish village

A Dreadful Visitation ~

THEY HAD BEEN CHILDREN THEN, they said, sixty or seventy years past. Many of their people, whom the white people later called Nez Perce, had gone east to hunt buffalo with their neighbors the Flatheads during the winter. In spring the families who had stayed behind crossed the mountains to join the hunters. They came to the hunting camp and found the lodges standing as usual, but their relatives whom they were expecting to greet were dead, almost all of them. Only here and there did they find anyone still alive. They took the survivors and returned home, but the disease followed them and killed them all, except some who ran away and a very few of those who stayed. In the beautiful Kamiah Valley beside the Clearwater River, their aged faces still bearing the marks of that long-ago spring, the missionary Asa Smith recorded their story in 1840. The scourge swept through the whole country, they told him, so that hardly anyone survived.

Though one was blind and the other two were very, very old, the three Coeur d'Alenes remembered what their fathers had told them about the terrible plague. They told the young Jesuit how their parents had carried the victims far away from their village on the St. Joe River and buried them on a rocky point, body after body, deep beneath the stones.

It had been about seventy years ago, they calculated, back when they were young. A few of their men had gone buffalo hunting when everyone left in camp fell ill. Large red pustules appeared on their bodies, especially their chests, and a few days later they died. On some people the pustules were black instead of red, and those

94

died almost instantly. Within a few days, only fifteen children were left alive in the entire village. One of them was the Flathead elder called Old Simon, who was said to be the oldest man in his tribe when he died in 1841. A few of the survivors were still living in the Bitterroot Valley in 1847 and passed their story to the pen of Father Mengarini. The same epidemic that visited their tribe, they said, had also appeared among another nation about five days' journey north, but there not a single person survived. "Of them remained not even the name," they told the priest.

→• ≡♦≡ •←

When fur trader John Work asked Colville elders in 1829 if their people were increasing or decreasing in population, they told him that immense numbers of their people had been swept off by a dreadful visitation fifty or sixty years before, and he noted as evidence the scars that marked their faces.

→• ≡♦≡ •←

Like the Nez Perce and the Flatheads, Kootenai bands living just west of the Divide often shuttled to the buffalo grounds on the northern Plains, and it was on such a spring expedition, the people said, that Charcoal Bull decided to raid a Shoshone camp. Several warriors entered a lodge and found a dead man inside. Although Charcoal Bull had strictly warned his men to leave everything alone, two young raiders secretly removed the dead man's moccasins. Some time later, as they returned home over a mountain pass just north of modern Glacier Park, one of the youths tried on the stolen moccasins. Both boys became sick, and their illness passed to the others. One of those afflicted was Bad Trail, whose sister dreamed that by roasting the fat of a buffalo's stomach and rubbing the grease on her brother's sores, the disease could be cured. She tried it, and her treatment was credited with saving eight of her brothers, the only survivors of their entire band. They were the ones who passed the story of the stolen moccasins to their children and their children's children.

＊＊ ≕◆≍ ＊＊

The two French Canadians journeyed west from the Saskatchewan River trade house with the Kootenais in the fall of 1800. For seventeen nights they traveled through the mountains, the twenty-six men and seven women, until they reached the Kootenai lodges. Later they moved south, to fine open plains. The Kootenais had but few horses, the Canadians said when they returned, but many were running wild as deer in the woods. "They have been in this state ever since the Time of the Small Pox in the Summer 1781," the travelers told Peter Fidler, "which swept away whole Nations."

＊＊ ≕◆≍ ＊＊

From a vantage point on a high knoll near the foothills of the Rockies, Piegan Blackfeet scouts looked down upon silent lodges in a camp of Snake Indians and decided to attack:

> With our sharp flat daggers and knives, [we] cut through the tents and entered for the fight; but our war whoop instantly stopt, our eyes were appalled with terror; there was no one to fight with but the dead and the dying, each a mass of corruption. We did not touch them, but left the tents, and held a council on what was to be done. We all thought the Bad Spirit had made himself the master of the camp and destroyed them. It was agreed to take some of the best of the tents, and any other plunder that was clean and good, which we did.

Soon the dreadful disease was spreading through the Piegan tents. An elderly Cree named Saukamappee who lived among the Blackfeet explained that "we had no belief that one Man could give it to another, any more than a wounded Man could give his wound to another." Saukamappee, narrating this story in 1787 to the Hudson's Bay Company apprentice David Thompson, recalled the people who rushed into the river for relief and drowned, and the shrieks and howlings of despair of those who survived.

⋯ ≕✦≔ ⋯

Hudson House was an early fur trading post situated just at the edge of the Prairies near present-day Prince Albert, Saskatchewan. In 1781 it represented the Hudson's Bay Company's westernmost outpost on the North Fork of the Saskatchewan River, their farthest push toward the Rocky Mountains and the unknown Columbia country. When clerk William Walker arrived in mid-October of that year to open the trade house for its winter business, he worried that nearby grass fires would make game scarce for the winter, and soon dispatched five men on a hunting expedition to the buffalo grounds. The group included a hunter named Mitchell Omen, who later told David Thompson that after the party had proceeded west for about 150 miles, they came to an Indian camp and saw people sitting on the beach beside the river as if to cool themselves. When they approached they saw that the natives were very weak and were marked with smallpox, from which they were just recovering. The furmen climbed the bank to the tribal camp and looked inside the tents. They were filled with victims of the disease. The survivors had moved their own tents a short distance away, too weak to go far, but a few of them had regained enough strength to hunt and keep the others alive. "They were in such a state of despair and despondence that they could hardly converse with us," Omen recalled. "From what we could learn, three-fifths had died under this disease."

While Omen's party was making this grim discovery, an Indian man had arrived at Hudson House who was taken with smallpox and said he had left seven of his companions dead in their tent on the plains. "This plaguey disorder," Walker wrote, "by what I can

97

hear was brought from the Snake Indians last summer, by the Different Tribes that trades about this River." Walker lay some of the blame for the pox's aggressive spread on the abundance of traded firearms and the subsequent increase of tribal raiding parties, which in his opinion brought about far more contact among disparate bands than in the past.

The next day three more infected natives arrived, and for the rest of the month workers were kept busy digging graves. Walker's journal described a pox that raged with such intensity "that in a short time I do not suppose there will hardly be a staid Indian living." The people apparently had begun to grasp the contagious nature of the disease, he noted, for many would not go near anyone who was stricken. This caused some who might otherwise have recovered to die of starvation—once afflicted, Walker said, they tended to give up any hope of recovery.

None of the company's European-born employees showed any symptoms of the pox, and although one of the mixed-blood workmen fell sick in early November, after fifteen bad days he was pronounced fit to return to work. But sick and dying Indians continued to struggle in, bearing reports of much worse devastation that they had left behind. Walker expressed shock and helplessness at their plight, and was concerned about how the tribes would be able to hunt and trap during the coming season—the disease was disrupting the established routines of the company. "It will be very detrimental to Our Affairs," he predicted.

On Sunday, December 2, after Walker read the Divine Service and his men buried one Indian woman, Mitchell Omen and his hunting party returned without any meat at all. Instead of buffalo, they had seen only more silent tents filled with dead bodies. The next day a man and two young boys came in, members of a group that had camped near the fur post in October. The smallpox had struck them on the day of their departure, and the three were all that remained out of five tents. Their band had taken a great number of skins and stored plenty of provisions the past summer, he told Walker, but now "the Wolves had destroyed it all, also pulled down the Tents and devoured the Bodies, shocking news indeed and so well we shall know it this Winter."

This was the situation that Walker tried to convey in a letter he sent on the fourth of December to William Tomison, his superior headquartered east beyond the Forks of the Saskatchewan River at Cumberland House: "I am sorry I should have such Disagreeable News to send You, but the smallpox is rageing all round Us with great Violence, sparing very few that take it." The few Indians in his neighborhood who had survived were frightened to go out "for fear of falling in with Others that is Bad." Hudson House depended heavily on traded provisions, and with little more than a month's worth of food left in the larder, Walker decided to ship half his men the 350 miles downstream to Cumberland House, with the confidence that Tomison could supply their needs.

William Tomison, however, was embroiled in a situation of his own. On December 11 a party had arrived at Cumberland House that included a woman "troubled with a Violent pain in her back & much inclined to Vomitting." Equally disturbing was news of more smallpox to the south. Tomison, the sort of man who counted hours and noted symptoms as they progressed, seemed more willing than Walker to stare the disease in the face. He checked regularly on the female victim, but his care did her no good, and she "expired between two & three in the afternoon being only the fourth Day of her ailment." At her subsequent burial, Tomison noted that her relatives would not touch the body.

A few days later more visitors arrived with reports of "that Devouring Disorder the small pox rageing amongst the Natives . . . & God knowes what will be the End thereof." Tomison had the new arrivals, including Walker's men, assemble all their belongings and smoke them with "the Flour of Sulpher" as a disinfectant. Tomison wrote that he traded with a group of Indians and "made them Presents as Usual but never expect to see them again."

When a boy was brought into the post with a violent pain in his breast and stomach, Tomison tended to him with fatherly attention. The second day his patient was still very ill, and on the third morning blisters erupted on his head and thighs. By the sixth day the lad could hardly swallow, but Tomison could think of nothing to help; "when Medicines come to be wanted, I am certain that there is nothing here to do us any Good." He assigned a man to watch over the boy and on the eighth night reported that the patient was now blind. Still he did not give up hope for the lad's recovery, worrying whether he would regain his sight after the disease passed. When a group of Indians arrived the next day who had not heard of the disorder, Tomison had a tent pitched for them some distance away and would not let them enter the post in hopes of preventing their infection. On January 4, 1782, the eleventh day of the Indian boy's affliction, one man tended the patient while the rest cut firewood and wove fishing nets. The next morning Tomison recorded "one man making a Coffin & one man digging a grave for the Indian lad he died last night between 9 & 10 OClock & was for 24 hours delirious."

As the winter wore on, more diseased and starving people staggered into the post. "Indeed their Condition is too shocking to be described by pen," Tomison wrote, as he dipped into his own scanty food supplies to feed a starving family. "I do assure your Honors, it cuts me to the Heart to see the Miserable condition they are in & not being able to Help." On the seventh of February, Tomison buried a close acquaintance named Wee'shen'now and learned that another had perished while trying to make his way to the post. These men were two of his key business associates, and he determined to try to salvage some furs to offset the value of the goods he had advanced them. He visited their tents, stripped the valuable beaver coats from a few of the bodies, and brought back seventy-eight beaver pelts. He buried four of the victims and pondered what to do about the thirteen starving people at the camp who were still alive. One had brought down a moose a day's journey away but was too weak to haul it in; Tomison sent some of his men to fetch the meat for the survivors.

Upon his return to Cumberland House, Tomison decided that he could not provide for all the men sent down by William Walker, and he ordered them back to Hudson House. It must have been a hard decision, and in a letter Tomison expressed sympathy with Walker's struggles. He also bemoaned their business losses, reckoning they had advanced supplies worth more than a thousand beaver pelts to the natives, which he now saw no way to recover. He was at a loss as to what to do about securing new transport canoes, "for all that Used to build are all Dead." Tomison could see that important knowledge and experience were being lost, and he fretted over the unknown fate of two dependable trappers named Sandfly and Pusas'quet'tumen. The next day, in what became a grisly ritual, he sent off men with duffle to use as burial shrouds, instructing them to bring in the furs of any deceased Indians they might find. The first crew returned with coats made from beaver and lynx, and reported that they had buried five Indians. Another pair of men discovered that wild animals had disfigured both bodies and furs.

Meanwhile, two weeks' journey upstream at Hudson House, William Walker indicated that the people around his post had apparently weathered the worst of the outbreak. After writing his December 4 letter to Tomison, he did not see one new case of smallpox until the first of February, and then no more again until the middle of March. But the post was still far from healthy; both whites and Indians battled hunger for the entire winter, and Walker's journal entries read as if the men had all lost their skill to hunt. He made some references to the scarcity of game, and others to women and children who had recovered from the smallpox but were starving to death. In a land traditionally rich in moose and close to the edge of buffalo range, the brigade stooped to snaring rabbits as their main source of food. The way the epidemic seriously disrupted the subtle rounds of winter hunting only underscored the company's dependence on tribal skills.

In the course of a normal year, the majority of men from outlying posts canoed downstream in May to Cumberland House, where they gathered the winter fur packs and ferried their take on to Hudson Bay for shipment to England. This year, however, William Walker decided to hold his own crew around Hudson House in order to comb the countryside for undelivered furs. In his last letter of the season to Tomison, Walker sounded almost optimistic that they could make something from the horrid winter after all. "There is a good few Indians alive up here yet, Some not over the Pox, and a great many young fellows gone to bring their own and the furrs of their deceased relations," he wrote.

As the winter tailed away, William Tomison kept his Cumberland House men at the nets, capturing sturgeon and pike in lieu of their preferred red meat. He ministered to the sick around him and ventured out in the field to learn the fate of Sandfly and his family. When the weather turned warmer, he fell with great relief into the more pleasant routines of spring, sending Indian women out for canoe gum and tracking the waves of migrant swans and ducks that would rejuvenate the house diet. Finally, in the middle of May, Tomison was forced by the shortage of hands to join the canoe brigade downriver to York Factory: "This great Misfortune Obliges me to go on the Head of the Journey Myself." But even then Tomison was not clear of the great misfortune; he soon found that the pox had preceded him to Hudson Bay.

The Mark ~

I BELONG TO THE GENERATIONS of humans who received vaccinations against the *Variola* virus. I can remember watching the steady prick, prick, pricking of the needle on my upper right arm, chosen because I already knew I was left-handed. I recall the small pustules that arose like dewdrops from the test circle, and how they itched like crazy. Just as my older brother had before me, I defied the admonitions of doctor and mother not to touch those pustules, picking at them until the clear liquid ran down, then scraping away the crusty shell that formed around the bull's-eye. Those fascinating pinpricks soon receded into little more than a curious shared birthmark, one that could be found on every other body splashing in the water at the beach. Sun and skin color and shape conspired to make each mark unique, and although my brother and I might stare at a particularly striking vaccination, we never connected them to the ancient dread they symbolized. Photos we saw in *Life* magazine of the last horrifying Asian cases of smallpox did not make the scabs on our upper arms tingle with empathy or fear. We had no idea that when we came down with the chicken pox, one after another, and got to lay out of school for a week, we were getting a very small glimpse of the Grim Reaper in full swing.

Before smallpox was eradicated, a case of the disease began much like chicken pox, with flu-like symptoms of muscle aches and fever that came on twelve to fourteen days after exposure. The pain could increase beyond imagination, often concentrating in one area, and sometimes led to severe anxiety or vomiting. After a couple of days, the fever backed off, only to return accompanied by the first characteristic red spots. These grew into blisters that were usually concentrated in the mouth and nasal passages. Within another day, a fast-moving rash began to run from back, neck, and forearms toward the palms of the hands and the soles of the feet. The denser the pustules, the poorer the prognosis. Mouth and throat irritation could lead to severe dehydration; cracked and runny sores invited secondary bacterial infection; swollen mucous membranes could cause blindness and sterility. Gradually a crust formed on top of each pustule, then hardened into a scab. Most deaths from smallpox occurred during the second week of the disease, but survivors had to endure the sores for twice that long, and they carried the visible scars for the rest of their lives.

The appearance of the first symptoms signaled that the patient was contagious. The disease was at its most communicable during the first week of the rash, but could still be transmitted as long as the scabs hung on—usually about a month after exposure. Infection occurred either directly, by breathing the airborne virus, or indirectly, by touching the oozing lymph or contaminated material and then inhaling viral particles hidden within them. Corpses remained contagious for up to three weeks after death; articles that the sick person had come into contact with, such as bedding or clothing, could remain infectious for more than a year.

Smallpox is an Old World disease, and through long association the peoples of Europe built up a strong resistance to its devastation. Outbreaks of the pox would roll through the crowded towns and villages every few years, hitting the very young and very old the hardest. Those who made it through had received a vaccination mark of lifelong immunity. Because smallpox was so embedded in European culture, it came to the New World as soon as the Spanish conquistadors began their rampage through the Caribbean and Mexico. From 1519 to 1521, the soldiers of Hernando Cortés kicked off an epidemic, which was described by their Franciscan missionaries. The scattered settlements of colonial America recorded outbreaks of greater and lesser impact over the next two and a half centuries. All observers noted the disproportionate effect the disease had on native populations, but few actually saw the first wave of infection. As horses and guns radiated ahead of Spanish, English, and French traders moving from the periphery of the continent into its interior, the frequency and speed of communications between the tribes increased dramatically. So did the prospects of a large-scale eruption of disease.

Initial reports of the Western Hemisphere's first clearly documented smallpox pandemic emerged from the Mexican plateau in the late 1770s. The disease swept south across the Guatemalan highlands and radiated north with equal vehemence; by 1780 it had appeared in New Mexican pueblos and across the southern Plains, homeland of the Comanche Indians. Since their acquisition of horses only a few decades earlier, Comanches often rode hundreds of miles north to trade with Shoshones on the eastern front of the Rockies and Crows on the northern Plains. In 1780 and 1781, smallpox could have passed from the Comanches to the Shoshones and then spread through the complex social web that crisscrossed the Columbia Plateau and the prairies. Both the Kootenais and the Piegan Blackfeet, occupying territory on either side of the Continental Divide, specifically pointed to the Shoshones (or Snakes, as they were often called) as the source of the outbreaks that afflicted their tribes. One tendril of the disease would naturally have flowed east from the Blackfeet along the Saskatchewan River to Hudson House and beyond.

Never Sits Down ~

JUST HOW FAR THE WESTERN arm of the outbreak reached from its nodes on the Columbia Plateau remains uncertain—the trade house journals of William Walker and William Tomison, literally hundreds of miles removed from the scene, represent the contemporary writing nearest to the Northwest. But beginning in 1787, sailors along the north Pacific coast reported the pockmarked faces of smallpox victims among tribes from the Queen Charlotte Islands south through Puget Sound to central Oregon. George Vancouver's expedition of the 1790s noted large empty villages, not all of which could be explained by seasonal movements. Lewis and Clark saw evidence of an epidemic among the Clatsops at the mouth of the Columbia, and on their return upriver in 1806 Clark noticed the ruins of a large village near the Willamette River. When he inquired as to the cause, an old man brought a woman forward to show her face, making signs to explain how she almost died when just a girl and how his people all had died. Clark put the old man's signs into words: "From the age of this woman this Distructive disorder I judge must have been about 28 or 30 years past, and about the time the Clatsops inform us that this disorder raged in their towns and distroyed their nation."

These sightings might have been part of the great 1780s pandemic, with the virus moving from the Plateau down the Columbia and then up and down the Pacific Coast. It is also possible that Vancouver, Lewis and Clark, and others saw the results of limited regional outbreaks that had moved south from Russian settlements on the Kamchatka Peninsula or arrived via a Spanish ship that traded in the Northwest in 1775. But wherever the smallpox came from, it was in the Northwest to stay.

Variola is a recurring virus; once established, it sweeps through succeeding generations. Thus smallpox reappeared in the Northwest in the early 1800s, just as the immune survivors of its previous visitation were raising a new population of susceptible targets. Again the disease came from the Great Plains, some said from the Crows, and again it swept across the Plateau, through the Flatheads and Nez Perce, to the Pend Oreilles and Kalispels, on to the Spokanes and the Colvilles. Some of the Plateau tribes remembered the second visitation as being less virulent than the first, but Clatsop people told Lewis and Clark that it had wiped out several hundred inhabitants and four chiefs only four years earlier. The captains described a number of deserted villages along the river and the coast as corroboration.

Memories of the scourge were still strong in 1811, when a Kootenai woman known as Qanqon dressed herself as a man and traveled down the Columbia prophesying disease and claiming to possess the "power to introduce the Small Pox." These prophesies so inflamed the Clatsops and Chinooks near the mouth of the river that the newly arrived Astorians took the supposed man inside their fort for protection. A few weeks later David Thompson also arrived at Astoria, where he recognized the Kootenai as the former wife of one of his voyageurs. Returning upriver with a group of Astorians and Qanqon, Thompson was approached by four fishermen at the Cascade rapids who inquired about "the Small Pox, of which a report had been raised, that it was coming with the white Men and that also 2 Men of enormous Size [were coming] to overturn the Ground." Noting that the men would gladly have plunged a dagger into Qanqon had he not stood in the way, Thompson assured them that there was no truth to this apocalyptic prophesy.

Duncan McDougall, chief factor at Astoria, might have taken a cue from the local people's strong reaction to Qanqon's boasts. When tension with the tribes reached a dangerous level a few weeks later, McDougall, according to his clerk Ross Cox,

> assembled several of the chieftains, and showing them a small bottle, declared that it contained the small-pox; that although his force was weak in number, he was strong in medicine; and that . . . he would open the bottle and send the small-pox among them. The chiefs strongly remonstrated against his doing so . . . that if the small-pox was once let out, it would run like fire among the good people as well as among the bad. . . . He was greatly dreaded by the Indians, who were fully impressed with the idea that he held their fate in his hands."

It soon became clear that the recurrent horror of smallpox was only one of many diseases that native people would endure—from influenza and malaria to measles, scarlet fever, typhoid, dysentery, tuberculosis, and diphtheria. Tribes that had numbered in the thousands became extinct, others mere remnants of their former strength. From the source lakes of the Columbia to its mouth, early anthropologists heard stories that involved single or paired survivors who wandered far, consumed with shock and grief, until they chanced upon one another. Together they provided the seed of a new village, and new life.

As decades passed, the fury of the sicknesses slowly abated. After more than two centuries, many of those diseases have become subjects of epidemiological history instead of dire prophecies. Smallpox, the precursor of them all, is now confined to isolated vials, tucked away in guarded vaults, awaiting either its own ordered extinction or a sudden call to protect unexposed humanity from another dreadful visitation.

⊷ �ईⅢ ⊶

Before the white men ever appeared in the Kootenai country, the story goes, when the people still cut their firewood with bone axes, Never Sits Down was camping beside a lake with his father and another man. The man became ill with the smallpox, and then Never Sits Down fell sick, too. Sores started breaking out on his body. When he saw what was happening to him, Never Sits Down walked out to the lake. The people in the camp had a belief that sore-ridden people should not touch water. To their amazement, Never Sits Down entered the lake and swam clear across it. He swam back, only to turn around and swim across to the far side again. Finally he disappeared from sight. Never Sits Down was gone for several days, and when he returned he was cured.

A Kootenai elder called Eustace told this story in 1930, surrounded by younger members of his St. Mary's band, near the same lake where Never Sits Down swam away his sores. Eustace said he had more stories about the exploits of the strong and resilient Never Sits Down, but before he would continue he wanted to make sure his listeners thought about what he had said. "This was a long time ago," Eustace explained, "but the blood of Never Sits Down is still around." He pointed to the man sitting beside him, then to others in the group; to many people. The "devouring disorder" has faded to a distant memory today, but the blood of Never Sits Down lives on.

CHAPTER SEVEN

Mount Coffin

Mount Coffin, after a watercolor by H. J. Warre, 1845

Calm Day ~

THE LOWER COLUMBIA has always worked as a place for composing pretty scenes. In a quiet watercolor executed during the first half of the nineteenth century, tribal dugouts traffic in both directions, and ducks preen and bow on a grounded log. There is no hint of a breeze; it is the kind of clear day and calm water rarely seen along this stretch of the river. On the far left, a rosy background glow rises to touch the snowy cone of a distant Mount Rainier, then continues across the horizon to envelop Mount St. Helens, much closer and smoking softly. Waves of green hills roll forward from the volcanoes, which are reflected in a circular white sheen that spreads across the river's surface. Behind that circle of light rises the focal point of the composition, a craggy-browed promontory that has captured the attention of both birds and people.

The top of the stack forms a round dome that is bald save for some stunted Douglas firs, and the downstream side of the mount drops almost vertically into a tangle of coniferous and deciduous trees. The opposite side descends more gently, so that the trees thicken before giving way to the rich riparian growth of coastal bottomlands. The whole formation floats like an island along the river's north shore, with water cutting behind it on both ends. A pleasant swath of green open space, dotted with standing fir snags, curves from the foot of the promontory to its summit. The scene is placid, the canoeists unhurried. Only the gnarled brown toes reaching down past the water's edge hint at an unquiet past.

Those toes were born of molten rock deep beneath the crust of the earth. The subduction of an oceanic plate beneath the western margin of North America around forty million years ago caused magma to stream up along undersea faults, forming giant coagulations of basalt, which slowly solidified and interfingered with shallow sediments near the shore. Uplifted into a coastal shelf around twenty million years ago, the Eocene basalts were overlaid by new waves of molten rock that crept down the Columbia, then reexposed when the course of the river cut back through the younger layers. As the landscape wore away, one shoulder of the original basalt stood firm. Rainfall and snowmelt from the slopes of Mount Rainier and Mount St. Helens joined to form the Cowlitz River, which flowed south to meet the mother Columbia. Between them, the two rivers sculpted the slopes of the promontory. Glacial floods crashing down the Columbia during the Pleistocene ripped the tops from hills around it. Violent lahars of pyroclastic mud from the slopes of St. Helens careened down the lower Cowlitz to pile acres of silt around its backside.

Soon after the last of the Lake Missoula floods inundated the monolith about thirteen thousand years ago, vegetation reestablished itself along the stack's accessible reaches. In the water around its toes, salmon, lampreys, and sturgeon resumed their seasonal runs upstream. Wapato plants sank their tuberous roots into the ashy silt of the floodplain. In time, Chinookan peoples constructed their lodges at favored sites around the isolated sentinel. They carved canoes from the big trees on shore and pulled fish from the river. They dug their bare toes into the marshes behind the rock to harvest nutritious tubers of wapato, then traded them upstream and down. When any of their people died, they wrapped the body well in furs or mats and hoisted a canoe atop the mount. They secured the body tight within, safe from rising tides and prowling beasts.

A Remarkable Knob ~

IN LATE OCTOBER 1792, Her Majesty's Ships *Discovery* and *Chatham*, under the command of George Vancouver, arrived outside the entrance to the Columbia River. The smaller *Chatham* took the lead across the raucous bar, but a freshening gale caught the *Discovery* in a precarious position. After three unsuccessful attempts at the crossing, Vancouver made for the open sea, leaving the *Chatham* and Lieutenant William Broughton to explore the expanse of water inside the breakers. Broughton's scant knowledge of the inlet came from a rough chart made by the American Robert Gray, who had crossed the bar only five months earlier in the *Columbia Rediviva.*

After repeatedly grounding on sandbars, Broughton anchored the *Chatham,* fitted out his launch and cutter with a week's provisions, and set off to find the source of the large "Fresh Water River" that emptied into the broad estuary from the east. The expedition moved at a surveyor's pace, with the crew fighting tides and current while Broughton recorded headings and sextant shots. He named prominent landmarks after countrymen or the first descriptive image that came to mind—Tongue Point, Pillar Rock, Puget Island. As the river narrowed, four Indian canoes fell in with Broughton's boats, and although neither party could understand a word of the other's language, the natives willingly sold fish to the Englishmen. Among the canoeists, Broughton recognized a "friendly Old Chief" who had visited the *Chatham* at the river's mouth. The next day the cutter's escort grew to nine canoes, whose paddlers won Broughton over with their "orderly behavior."

On their second day inland, as the river began a long bend to the south, Broughton recorded on the north bank "a remarkable mount, about which were placed several canoes, containing dead bodies; to this was given the name of Mount Coffin." About a mile beyond this landmark they stopped at a lodge beside a stream, which the Englishmen christened Hut Creek. The next morning the party passed a smaller, rocky islet, sticking twenty feet above the surface of the water, upon which were perched the same sort of burial canoes. On his chart Broughton labeled this uplift Burial Head; succeeding visitors renamed it Coffin Rock.

The explorations of Robert Gray and Lieutenant Broughton uncorked the mouth of the Columbia River, and in succeeding

years ships from Boston and England occasionally braved the bar
to trade. The sailors never ventured beyond the wide protected
bays at the river's mouth, but in time new terms like *dag, ax,
damned rascal,* and *son of a bitch* rippled upstream and bounced
between campfires at the foot of Mount Coffin. The clank of tin
powder flasks and the occasional bang of a musket rose from the
deciduous bottomlands. Glass beads hung beside dentalia shells
on the mount's burial canoes, and striped blankets lay beneath
rush mats. Piped sailor jackets appeared beside cedar bark skirts in
the canoes that passed beneath its brow. And within a little more
than a decade of Broughton's visit, new visitors approached Mount
Coffin's steep nose from the direction of the sunrise.

"I had like to have forgotten a verry remarkable Knob riseing
from the edge of the water," William Clark wrote at the end of his
journal entry for November 6, 1805. The party had floated past the
knob on the opposite side of the river, where two canoes emerged
from nearby lodges to trade fish and wapato roots. Clark identified
these men as members of the "Scil-loot Nation," whose main villages clustered around the mouth of the Cowlitz River. They spoke
a dialect of the Chinookan language prevalent along the lower
Columbia, and, like many of their brethren, used wooden cradle-
boards to flatten the heads of their children. During the winter at
Fort Clatsop, several members of the tribe visited the Americans,
bringing tobacco and wapato to trade on one occasion and a gun to
be repaired on another. Lewis changed the spelling of their name
to Skil-lute and concluded that these people were the primary
traders between the sea coast and tribes upriver as far as the Dalles.

When the explorers returned upstream in late March 1806, they
camped in the shadow of the "remarkable high rocky Nole." Next
morning, at a nearby Skillute village, they breakfasted on eulachon
and sturgeon, wapato and camas roots. Lewis estimated the Skillute
population at around thirteen hundred and noted that their language differed slightly from other Chinookan bands along the river.

The fur trade arrived in earnest on the Columbia in 1811 when the Pacific Fur Company established a post at Astoria, and soon Canadian and American furmen began to work the entire length of the river. French-Canadian voyageurs bestowed the title of *Mont des Morts* (Mount of Death) on Clark's remarkable knob, and rare was the passing diarist who failed to mention it. Some scribes marveled at the stack's noble and isolated stance; others made wildly varying estimates of its elevation. But most of all, they saw the mount as a focal point for local funerary practices, "a receptacle for the dead."

Congregated Dust ~

FURMEN GAWKED at the sheer number of the canoes on Mount Coffin: "All over this rock—top, sides, and bottom—were placed canoes of all sorts and sizes." Many shivered at the gothic image of vessels "containing relics of the dead, the congregated dust of many ages . . . this sepulchral rock has a ghastly appearance, in the middle of the stream, and we rowed by it in silence." Some stopped to picnic on its upstream flank and explore the cemetery; one trader measured a burial canoe, fashioned from a single log, at five feet broad and four feet high at the stern. Nearby lay a large seagoing dugout decorated outside and in with seashells of various kinds.

Canoe burial served as the main form of interment for Chinookan bands from the Willamette River to the Pacific, and along the coast between Willapa and Tillamook Bays. When naturalist John Kirk Townsend explored Mount Coffin in 1835, he found bodies carefully wrapped in blankets, with the personal property of the deceased—bows and arrows, guns, salmon spears, various ornaments—placed inside and around the canoes. Holes were bored in the bottom of the dugouts to keep water from collecting inside, and they were covered with mats to protect them from the constant drip of coastal weather. Wealthier families topped the burial vessel with a larger canoe turned upside down and lashed securely.

Modeste Demers, a Jesuit priest who lived on the Cowlitz
River in the late 1830s, observed many burials; he saw eyes band-
aged with chains of glass beads, and nostrils filled with small white
dentalia shells that were imported from Vancouver Island and
highly valued. Each burial canoe was positioned to point downstream, he said, and each body laid face down inside. The weapons
of the men, and the digging sticks of the women, were placed at
their sides; lesser objects were hung on poles around the boat.
Many observers puzzled over the common practice of "killing"
dishes and kettles by puncturing them with awls before arranging
them around the burial site. Some thought the utensils were
placed there to be used in an afterlife, but George Gibbs, who
interviewed Chinook elders in the 1850s, recorded their belief that
the deceased visited their grave sites every night and "would be very
angry if they found their property in use by others. . . . They value
these things very much, and come to look after them."

Furmen who lived among the tribes heard songs of mourning
chanted at dawn and dusk and watched wives, relatives, and slaves
cut their hair and visit a cemetery twice a day for some time after a
loved one's death. One early traveler remarked that the natives
would not eat or laugh in sight of the burial canoes "for fear their
mouths will be turned askew" by their departed family members,
who they believed were still aware of what went on among the living. Elders described two countries of the dead, called *memaloost
illahie*, to George Gibbs. Souls remained in the first until their
flesh and bones had entirely disappeared, during which time they
revisited the earth at night and could be seen by men. Sometimes
the dead came to their friends in dreams. After their bodies entirely
decomposed, they crossed a river into a second country and were
gone forever. "The road to the land of the dead is by the West, for
which reason the head of the corpse is placed toward the setting
sun," they explained.

No matter how fragmented or incomplete these outsiders' views of tribal religion might have been, one constant in all accounts was the great respect the natives exhibited toward their ancestors. John Kirk Townsend, at anchor in a brig near Mount Coffin, wrote that "the vicinity of this, and all other cemeteries, is held so sacred by the Indians, that they never approach it, except to make similar deposits." He had seen tribal people make long detours to "avoid intruding upon the sanctuary of their dead." And like many of his predecessors, Townsend saw no contradiction in visiting the "tombs" on Mount Coffin himself. As he and his companions walked around the site, they were careful not to touch or rumple any of the shrouds,

> and it was well we were so, for as we turned to leave the place, we found we had been narrowly watched by about twenty Indians, whom we had not seen when we landed from our boat. After we embarked, we observed an old withered crone with a long stick or wand in her hand, who approached, and walked over the ground which we had defiled with our sacreligious tread, waving her enchanted rod over the mouldering bones, as if to purify the atmosphere around, and exorcise the evil spirits which we had called up.

Sacred Mementos ~

LIEUTENANT BROUGHTON AND HIS CREW had paused on their survey of the Columbia to visit the cemetery at either Mount Coffin or Coffin Rock because, as the master of the *Chatham* later wrote, "Curiosity led them to it, and it proved a Receptacle for the Dead: the bodies were wrapped in Deer and Bear Skins: and like the custom of De Fuga Indians had their warlike weapons with them." Curiosity also led Broughton to examine a grave near the mouth of the river, but he made sure that everything he had displaced was "restored to its original situation." Many who followed did not show such restraint, collecting burial artifacts to ship back home as

downriver. This time his crew was made up of Hawaiian paddlers who were not so observant of local taboos, and Scouler persuaded them to pull ashore on Mount Coffin. There he walked among vessels decorated with bird feathers, laid with carved wooden bowls, and picketed by "boards painted with rude resemblances of the human figure." All of them were situated along a steep slope of the rock that faced the river, and it was Scouler's opinion that the very steepness of the site controlled the number of canoes on the mount at any one time, because as they decayed they naturally rolled into the river to be carried out to sea. The burial canoes were covered with boards, then sealed with strong cords and weighted with large stones, and Scouler found the protective planks so securely bound that he couldn't see what lay within. Not wishing to damage any of the wrappings, he searched until he found an older canoe that was well decayed; when he lifted one of the boards, a snake wriggled out into the night. Behind the serpent lay a complete skeleton festooned with the usual burial ornaments. Working around these trappings, the surgeon availed himself of "the opportunity to procure a specimen of their compressed skulls."

Even as Scouler violated the sanctuary of Mount Coffin, he was moved by the site. "This method of burying the dead, if I may use the expression, is very affecting. The solitude of the place & the assemblage of so many objects with which we are not accustomed to associate serious ideas, deposited as mementos of the dead, can not but form an interesting contrast & give rise to the most serious reflections." The doctor did not hesitate to visit the Skillute village for breakfast the next morning, where he was treated with great civility. Before leaving the Columbia, Scouler secured two other skulls; back in Scotland he made a series of scientific measurements on them and published a journal article in which he chastised naturalists for their obsession with mosses and zoophytes while ignoring the natural history of the human species. With the intention of replacing "wild theory" with careful investigation, he presented a description of the compression of childrens' heads by Northwest tribes, concluding that "at all events, it has no effect on their intellectual powers."

Meanwhile, tribesmen on the Columbia had discovered his thefts, and even though a fellow scientist justified his actions as "robbery for the sake of science," Scouler's friend David Douglas was so struck by local reaction to the incident that he clarified his own collecting limits in a letter to a mutual friend:

> I intend to procure the skulls of dogs, wolves, and bears for Scouler, but none of men, for fear he should make a second voyage to the Northwest coast and find mine bleached in some canoe, "because I stole from the dead," as my old friends on the Columbia would say.

Scouler found a less hesitant supplier in William Tolmie, another physician employed by the Hudson's Bay Company. During his years on the Columbia, Dr. Tolmie regularly boxed up natural and tribal artifacts for his contacts in England. One shipment to a Hudson's Bay Company governor contained a single "cranium"; a larger package to Sir George Simpson held six curious shells and seven flattened heads. Two crates addressed to Dr. John Scouler were packed with tribal pipes, dishes, masks, and bird skins, plus three more skulls.

John Townsend was disarmingly honest about the subject, writing that he was "very anxious to collect the skulls of some of these Indians" but had checked the impulse on his first visit to Mount Coffin because he did not wish to endanger his companions. Later that year, acting alone, Townsend did manage to steal several skulls. He used utmost caution to leave the canoe coverings exactly as he found them so no one would notice the theft, but confessed misgivings about his activities: "I thought several times today, as I have often done in similar situations before: Now suppose an

Indian were to step in here, and see me groping among the bones of his fathers, and laying unhallowed hands upon the mouldering remains of his people, what should I say?"

When Townsend returned to Philadelphia, he gave the skulls he had collected to Samuel George Morton, a physician with a special interest in comparative human phrenology. When Morton published his *Crania Americana* in 1839, comparing skulls of various aboriginal peoples, he included measurements, descriptions, and detailed drawings of eight skulls from the lower Columbia River, from individuals ranging from slaves to chiefs, as well as a fine example of a skull-flattening cradleboard. Morton's studies led him to conclude that the American natives were of one race and one species, countering earlier scientists who had argued otherwise.

But the advancement of science was little consolation to those whose cemeteries had been robbed, a capital offense among the Chinook people. A British sea captain noted in 1839 that the tribes had become very secretive concerning their burial ceremonies, partly from fear of prying Europeans, who were stealing not only skulls but grave ornaments as well. Oregon pioneer Tolbert Carter watched a bateau captain, whom he characterized as "somewhat of a curio hunter," strip valuables from a body on Coffin Rock in 1847. Disgusted by the sacrilege, Carter wrote, "What made the scene more impressive to me was that the time might come when our race would become extinct, and our own bones disinterred by the living race to find curios of the people that once existed." This was a sentiment shared by fur agent James Swan, who lived among coastal Chinooks in the early 1850s. "They regard these canoes precisely as we regard coffins," wrote Swan, "and would no more think of using one than we should of using our own grave-yard relics; and it is, in their view, as much of a desecration for a white man to meddle or interfere with these sacred mementoes as it would be to us to have an Indian break open the graves of our relatives."

The Fever Ghoul ~

IN 1829 THE SHIP OWYHEE entered the Columbia River to trade and fish commercially for salmon. Captain Dominis anchored near Deer Island, just upstream from Mount Coffin, and dispatched trading parties up and down the river. A teenaged apprentice aboard the *Owyhee* later narrated to a historian his memories of the year he spent on the Columbia, recalling a sickness that broke out among the Indians. "It seemed to originate with the Indians about the ship," he said. "The disease became epidemical and malignant, so that whole villages died, and there were not enough well persons to care for the sick." Apparently the first victims were some mischievous boys who had been pulling up stakes set in the river by the Yankee fishermen, and the apprentice said the Indians believed that the disease was "intentionally brought about by Captain Dominis, who, they said, had emptied a vial of bad 'medicine' into the Columbia River with the design to destroy them." The *Owyhee's* crew would probably have suffered reprisals, he felt, but for the influence of John McLoughlin, chief agent of the Hudson's Bay Company at Fort Vancouver.

In late September 1830, McLoughlin, who was a physician, wrote from Fort Vancouver that "the Intermittent Fever is making a dreadful havoc among the Natives and at this place half of our people are laid up with it." Two weeks later he counted fifty-two on the sick list and estimated that it had carried off three-quarters of the Indian population in the vicinity. He treated sufferers with quinine until exhausting his supply, then turned to dogwood root as a substitute. Ascending the river that same month, David Douglas wrote:

> A dreadfully fatal intermittent fever broke out in the lower parts
> of this river about eleven weeks ago, which has depopulated the
> country. Villages, which had afforded from one to two hundred

effective warriors, are totally gone; not a soul remains. The houses are empty and flocks of famished dogs are howling about, while the dead bodies lie strewn in every direction on the sands of the river.

Douglas counted himself lucky not to have contracted the fever himself, and when he arrived at Fort Vancouver he learned he was one of the few people to remain well. Toward the end of November the pestilence broke out with increased vigor, and McLoughlin's sick list swelled to seventy-five. But while European and mixed-blood furmen usually recovered from the illness, natives often did not. Whole villages came upstream to camp near Fort Vancouver, "giving as a reason that if they died they knew we would bury them. Most reluctantly on our part," McLoughlin wrote, "we were obliged to drive them away."

Peter Skene Ogden, one of McLoughlin's traders, visited afflicted villages downstream from Fort Vancouver, "where the fever ghoul has wreaked his most dire vengeance." Ogden had heard of a similar disease in California, and believed it not to be contagious, but to come from "miasma pervading the atmosphere" and "foul exhalations from low and humid situations." Many historians see truth in Ogden's assumptions, for the range of symptoms described resembles malaria, which can be carried by a species of *Anopheles* mosquito indigenous to the lower Columbia. If sailors on board the *Owyhee* or any other vessel on the river that year were infected with malaria, they could have introduced a new plague to the region. Once malaria is established in the mosquito population it becomes endemic, returning every summer, and that is what happened on the Columbia. The malady, known as intermittent fever, *fièvre tremblante*, ague and fever, and the cold sick, was still in evidence four years later when John Townsend arrived to give the first clinical description of the disease. "The symptoms are a general coldness, soreness, and stillness of the limbs and body, with violent tertial ague," he wrote. "Its fatal termination is attributable to its tendancy to attack the liver, which is generally affected in a few days after the first symptoms are developed." Although simple tonic remedies prevented many deaths at the fort, it was impossible to treat all the far-flung natives. "The aspect of things is very melancholy," he remarked.

When William Tolmie ascended the river in 1833 to help fight this continuing epidemic, he floated past numerous burial canoes on Mount Coffin and Coffin Rock; two years later a missionary noted that Mount Coffin was covered with canoes. British captain Edward Belcher stopped at the mount in 1839 and found it "studded, not only with canoes, but, at the period of our visit, the skulls and skeletons were strewed about it in all directions." A Jesuit priest traveling upriver in 1842 stopped opposite the mouth of the Cowlitz, near a camp of Indians who spent the night singing and stamping their feet in an attempt to free themselves from an attack of "trembling fever." Dr. Tolmie spoke of so many deaths that they grew beyond the ability of relatives to keep up with the burials, and of survivors deserting some of the areas hardest hit. The intermittent fever, he wrote, "has almost depopulated Columbia R. of the aborigines." Anthropologists, who estimate the decline in native people on the lower river at approximately 90 percent between contact and 1850, believe that around this time the surviving Skillutes moved away from the mouth of the Cowlitz; their last known village was located at Oak Point, downstream from Mount Coffin. Within two decades that village too had disappeared.

A Supper Fire ~

IN MAY 1841, LIEUTENANT CHARLES WILKES, commander of the U.S. Exploring Expedition, passed the foot of Mount Coffin. Wilkes was immediately taken with the scene, commenting that the great numbers of canoes that littered its slopes in every direction "were as fast going to decay as the living." He also recognized that the landmark afforded a favorable point for observation and returned in late August with a surveying crew. As the men carried their instruments to the top of the mount, they passed hundreds of canoes in all stages of decay, many supported by tree limbs four or five feet off the ground. Beneath a clear and beautiful sky, the surveyors spent the day on the summit, taking a full set of observations on signal flags they had posted up and down the river.

After supper at the bottom of the mount, Wilkes shoved off with his crew to return to his ship. Apparently someone carelessly forgot to extinguish the cooking fire, and as the party rowed away they noticed that the flames had gotten out of control. "I regretted to see," wrote Wilkes, "that the fire had spread and was enveloping the whole area of the mount; but there was no help for it. The flames continued to rage throughout the night until all was burnt." Next morning the lieutenant presented the local Indians with a few small presents and explained that the fire had been an accident. While Wilkes said that the people appeared satisfied with his explanation, he conjectured that only a few years earlier—before the strength of the tribes had been decimated by intermittent fever—"the consequence of such carelessness would have been a hostile attack."

Wilkes's supposition was confirmed when Canadian artist Paul Kane visited Mount Coffin five years later and learned that the event had not been forgotten. "Commodore Wilkes having made a fire near the spot, it communicated to the bodies, and nearly the whole of them were consumed," he wrote. "The Indians showed much indignation at the violation of a place which was held so sacred by them, and would no doubt have sought revenge had they felt themselves strong enough to do so." Yet in spite of this bitter incident, the people must have continued to use the mount as a cemetery, for Kane estimated that two or three hundred canoes were deposited there at the time of his visit.

In September 1845 British army lieutenant Henry James Warre canoed down the Columbia on a mission of gathering information related to the coming boundary settlement between Great Britain and the United States. Warre, who had traveled across the continent from Montreal and seen his share of pleasant scenery, wrote that the section between Fort Vancouver and Astoria offered "not much to attract the Eye or worthy of notice." Then, almost in spite of himself, he began to describe the pretty views the day afforded — the rolling hills and greenery, always broken by the cone of Mount St. Helens, "standing as it were, at the head of the bend!" The mountain was in an active phase just then, and a long black column of smoke and ash shot into the air, then settled over the mountain's snowcap, managing to stir even Warre's curiosity.

His party lingered just below the confluence of the Columbia and Cowlitz Rivers, opposite the "peculiar feature" that his paddlers told him they called Mount Coffin. Warre wrote that he found the slopes "literally covered with Canoes containing the remains of departed Indians." He ran through the catalog of first impressions that many other visitors had used to describe the burial ground, noting torn blankets hung on poles for streamers at one fresh tomb.

From a vantage point on an island in the river, Warre took out his painting supplies and made a watercolor sketch of Mount Coffin. If you compare the sketch with a more finished painting that he must have executed at a later date, you can see the way he carefully lined up his composition to take advantage of the distant cap of Mount Rainier and the closer cone of Mount St. Helens. A steamy cloud rises from the left flank of St. Helens, separating it from Mount Coffin. This sketch emphasizes the path of the fire left behind by Lieutenant Wilkes's survey crew only four years previous, with charred snags tracing a raw scar across the face of the open slope. Where the low, curved outcrop meets the water, surviving Douglas firs form a thick ragged stand. In the first study, there are no people on the river, only dark shadows that leak out from Mount Coffin's downstream flank.

In the more formal painting, shades of pink and amber soften the tone toward evening, and the steam cloud from Mount St. Helens spreads back against the slope of the volcano. Warre removed several of the ghostly snags from the slope of Mount Coffin and chose to green up the rest. He replaced the copse of firs on the stack's right flank with a pillowed canopy of deciduous trees. He lightened the burn path and the black shadows on the left, and added a diagonal succession of tribal dugouts and loafing birds to the river. Although Warre was not a delicate enough painter to render animal shapes or human expressions believably, these changes do add a measure of life to the scene. But nowhere, in either version, did the artist include any hint of the rotting canoes or burial ornaments that signified to generations of people the essence of Mount Coffin.

Stars and Cinders ~

In November 1846 the American ship *Toulon* crossed the Columbia bar and sailed upriver, bearing news that the United States and Great Britain had signed a treaty establishing the international boundary at 49 degrees North. This was bad news for the Hudson's Bay Company but good news for American settlers, who had been clustering south of the Columbia on the assumption that the river would form the border. As waves of immigrants continued to arrive from the States, attention turned to the lands lying north of the Columbia, and the area around the Cowlitz River received particular notice. In the winter of 1847, pioneer P. W. Crawford canoed down to the mouth of the Cowlitz with a friend to stake a claim to a large parcel of land, which he accomplished by branding his name on a white fir beside the river. Continuing down the Columbia, they stopped at the high promontory of Mount Coffin and ascended the terraces along its south side. The view from the top was grand, Crawford later remembered, with the mottled islands of the Columbia to the south and a large prairie to the north, interlaced with sloughs that were skirted with willow and ash. "Quite a picturesque appearance," he recalled. But nowhere that day in all his ramblings did he see a lodge or a village. After the departure of the Skillutes from the area, Cowlitz bands descended the river seasonally to fish and gather roots around Mount Coffin, and a small group took up residence on an island at the mouth of the Cowlitz. But they too continued to suffer from fever and ague, and their numbers were also on the decline.

The establishment of Oregon Territory in 1848 included the lands north of the Columbia, and the Donation Land Law of 1850 formalized the process by which land could be obtained in the region, offering 320 free acres to any male American citizen over eighteen years of age who settled in Oregon Territory, with another 320 for his wife if he were married. In 1851 the government began negotiating treaties with the lower Columbia tribes regarding dispensation of their lands, but none of them were ever ratified. Around the mouth of the Cowlitz, there were few natives left to protest as claim stakes were planted on both sides of Mount Coffin.

A man named Noyes Stone laid claim to a piece of land behind the promontory, and when surveyors later platted township and range demarcations, the rocky knob lay precisely on the edge of the property line. A frustrated gold prospector named Crumline LaDu came up from California and settled on the adjacent parcel just downstream from Mount Coffin in 1851, then began farming strawberries in the rich bottoms off its flank. When steamships began running between Astoria and the burgeoning city of Portland, Mount Coffin marked the halfway point of the route.

Deep water at the base of the rock made good moorage for boats, so LaDu built a dock beside the mount that became known as Steamboat Landing.

In 1863 a young man named Daniel Webster Bush, who had married LaDu's daughter Alice, purchased the section of land that included Mount Coffin; he and his bride built a house a short distance upstream and began raising a family that eventually included thirteen children. Perhaps the arrival of those children created the need for the long picket fence shown in an old photograph stretching from the front of the homestead across a driftwood-strewn shore to the knobby toe of the landmark. The Bush children did not need to cross that fence to climb to the top of the promontory, which became a playground for many neighboring youngsters. They filled jars with beads and arrowheads and when they were older spooned with lovers on a certain round rock in a laurel grove on the broad summit.

In 1873, Crumline LaDu obtained a contract to start a post office, with "Mount Coffin, Washington Territory" as postmark. Bush erected a warehouse, and the landing became a shipping point for farmers and a convenient embarkation spot for local passengers. As a picturesque stop for steamboats, the mount provided regional publications with regular fodder; an etching in an 1886 magazine showed Mount Coffin in all its glory, rising above an array of steamers and sailing vessels whose wakes mark the glassy river. The Bush fruit warehouse, tiny by comparison, nestles beside a grove of leafy alders and cottonwoods. A line of stately Douglas firs climbs the slope to culminate in two large sentinels on the flat summit.

In time a second and third generation of settlers' children played on the rocky slopes, and new chapters were added to the mount's lore. During the great flood of 1894, rising waters forced the Bush family to move everything they owned to the second floor of their house, but their piano wouldn't fit up the stairs. They lashed it to the living room ceiling and watched the swirling river inch toward its legs, then recede. A granddaughter who waited out more than one spring freshet at the Bush homestead remembered that they always took comfort from the sight of Mount Coffin nearby, for they knew they would be safe there from any flood.

In 1908, Daniel and Alice Bush celebrated their golden anniversary in their home below the mount. That same year, possibly because Alice's failing health necessitated a move to Portland, Bush sold Mount Coffin to Star Sand and Gravel of Portland for eleven thousand dollars. The company began blasting out boulders and set up a rock crusher to systematically grind the ancient basalt into gravel. Barges replaced steamboats at the landing, taking on material sized for dikes and jetties, foundations and road fill. Thousands of tons of riprap were floated downstream to create Fort Stevens and build the lighthouse at the mouth of the Columbia. Thousands more tons of fine gravel were barged upstream to pave the new streets of Portland.

When the sawmill town of Longview bloomed into existence, there was a movement to save what remained of the area's signature landmark. The Long-Bell Lumber Company made an offer to purchase the land for a park, but Star Gravel turned them down. In 1927, with about a third of the promontory chiseled away, the local Daughters of the American Revolution attempted to secure it as a heritage site. The deal never came together, and a photograph made the following year shows a lumber warehouse fronting the familiar shape of Mount Coffin, with the wide, island-specked Columbia behind. The mount has been stripped of trees save for one leaning snag right on top, and the mining operation has taken a huge bite from its belly.

As the offices of the Weyerhaeuser paper mill grew up on the site of the old Bush homestead, large construction projects in the 1930s and 1940s demanded great quantities of rock, and the remaining crescent of Mount Coffin was gnawed away. By 1945 only a single crag of Eocene basalt was left. The *Portland*

Oregonian ran an obituary for the historic site, and a Bush granddaughter commented that her childhood haunt had a long past but a short future. In 1954 only one short knee of basalt remained beside the river, topped by a single stubborn fir. This small knob, not very remarkable at all, was finally leveled to make way for a cargo dock beside the lumber mill.

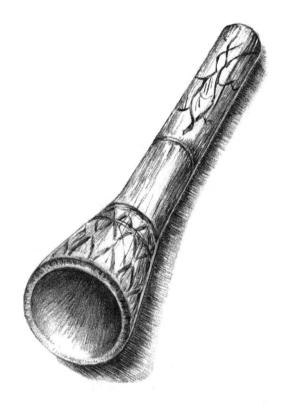

CHAPTER EIGHT

Smoke

Moses Coulee pipe case

Time's Garden ~

I WANDERED IN A GARDEN. It occupied considerable ground, surrounding the irregular bends of an old convent and changing tone and aspect around each hidden corner. Plantings tucked close to the main building were shaded by large exotic trees, and with dusk closing quickly, many of the shrubs and perennials seemed to reach beyond their natural limits. A narrow walkway skirted the sawn trunks of three huge maple trees, fronted by two irregular beds where gardeners had planted ornamental annuals. There I came upon the deep green elliptical leaves of a short nicotiana, with sprays of pleasant blossoms that opened around my shins. They carried a look that has long been popular in the Northwest; in fact, seeds from similar nicotianas were gathered alongside those of common petunias at a recent archaeological dig on the Ingles family homestead near the old Fort Vancouver fur post. It is easy to see why Mrs. Ingles would have favored the ornamental tobacco. On this nippy September evening, their trumpet flowers still bloomed in profusion, and the pale yellow and cream petals deftly captured the graying light. The delicate fragrance of their night-scenting blooms mixed with the pleasant rot of garden compost to curl through the air.

A sphinx moth helicoptered over and touched two or three of the blossoms, nectaring so quickly that all I could make out was the urgent buzz of its wings and a dark bouncing shadow that zipped across the footpath and disappeared past the overgrown beds. In its wake I noticed another variety of nicotiana, this one head-high and much more robust. Beneath flowers as white as the moon, the stems grew thick and hairy, and the leaves spread larger than my outstretched hand. One touch confirmed that the green foliage was covered with sticky brown hairs, each one a gland bursting with the oil of nicotine. This was a rough-cut cultivar, one whose acrid, almost chemical smell overpowered the sweet nectar of its cousin's blooms.

I grew up in country where tobacco was a profitable crop, at a time when smoking permeated the culture, and these stalks brought back the memory of flue-cured leaves hanging rich and golden on bare pine poles. I could see the yellow-green sprouts of the commercial fields in the Carolinas, and remember how children came crying out of them in late summer, their fingers swollen by those prickly glands of nicotine. I could smell the inside of my Uncle Mutt's Pontiac, saturated with the noxious, sweet aroma of the perpetual cigar that lolled from one side of his mouth to the other.

Points of Etiquette ~

As THE THREE CARAVELS of Christopher Columbus set sail from the newly christened island of San Salvador in 1492, they were hailed by a lone boatman in a dugout canoe. The man scrambled aboard the *Santa Maria* for a ride to the next island, carrying with him a clump of red earth and "some dry leaves which must be something much valued by them, since they offered me some at San Salvador as a gift." When Columbus landed in Cuba a couple of weeks later, he saw Taino people rolling similar dried leaves into long cigars and sniffing smoke from the "tobacos" through their nostrils.

The cured herb of the Tainos belonged to the genus *Nicotiana*, whose ancestral forms arose in South America and spread along the backbone of the American continents. Except for a few representatives in Australia and the nearby South Pacific Islands, the members of this genus grow only in the Americas, where they tend to favor rather open, semiarid country. Some of these tobacco species live as annuals, others as perennials; some hug the equator, while others flourish in high temperate latitudes. They range in size from a scraggly bush to an actual tree, with wide variations in form; they hold in common trumpet-shaped flowers and prickly hairlike glands. At some point in the distant past, humans discovered the potent pharmaceutical properties of tobacco chemistry, then learned how to propagate the plants from seed. Early travelers introduced a couple of big-leafed South American species across the Caribbean islands and up through Mexico and eastern North America. Species native to western North America were also spread along aboriginal trade routes, their tiny seeds moving from tribe to tribe, hand to hand.

Beginning with Columbus, and continuing with Cortés in Mexico, Cartier in the St. Lawrence, and Drake in California, explorers around the periphery of North America watched tribal people imbibe tobacco. When sailors carried sample plants and seeds back across the Atlantic, fashionable people in Europe began to smoke, chew, dip, snort, and sneeze themselves. French ambassador Jean Nicot brought the plant to the attention of classifying scientists, who attached the diplomat's name to the poisonous alkaloid that humans had bent to their use. By the early 1600s, cultivated varieties of *Nicotiana tabacum* were being grown on a commercial scale in Brazil, Jamaica, and Virginia. The wide leaves were air-cured, then spun into rolls called carrots, or formed into the wrapped loops known as twist.

As explorers and fur traders worked their way across the interior of North America, they carried manufactured tobacco for their own use, and soon discovered that a few inches of twist served as a salve to grease civil transactions with the tribes they met, many of whom already had long-established rituals of smoking.

In the summer of 1787, naturalist Archibald Menzies stepped ashore on the Queen Charlotte Islands and collected samples of a peculiar tobacco plant that was being cultivated in the garden of a Haida man. Menzies pressed his botanical specimen and sailed away, leaving it to two American sailors to document a practice among the Haida and their Tlingit neighbors of pounding green tobacco leaves together with limestone, then placing quids of this mixture "as big as a hen's egg" into their mouths. One of the Americans described a fine meadow covered with white and red clover, wild celery, and some tobacco plants; the other collected seeds to send to a friend. Neither suspected that the Queen Charlottes lay hundreds of miles north of any known range for wild tobacco.

During his initial survey of the lower Columbia River in 1792, Lieutenant William Broughton noted that the tribes there smoked an herb that grew in the area through long-stemmed pipes that featured carved wooden bowls attached to elderberry tubes, often two feet in length. A crew member commented that a family shared one pipe, each person taking a hearty suck and then swallowing the smoke. Broughton distributed some of his imported tobacco, "which seemed to give the people much pleasure."

A few months later, a group of Kootenai Indians from the opposite end of the Columbia became the first known members of a Plateau tribe to share a pipe with a white man when they journeyed east across the Rockies to trade horses with the Piegan Blackfeet. Traveling with the Blackfeet was Peter Fidler, a Hudson's Bay Company agent who recorded the encounter in great detail in his journal. The parley between the Kootenais and the Blackfeet began with an introductory smoke that lasted three-quarters of an hour. After bartering a few horses and skins, the Kootenai chief lit a pipe and made a speech in his own tongue, of which Fidler could not understand a single word. The Kootenai then took four whiffs of the pipe and handed it to Fidler, who was smoking away at his ease when the chief retrieved the pipe, indicating by signs that four whiffs was the proper number to take upon significant occasions such as making peace, or meeting friends and strangers.

When that pipe was exhausted, the chief refilled it with "his tobacco, all of their own Growing." He presented it to Fidler and signaled that the furman should now demonstrate the white people's manner of greeting and smoking. Fidler obligingly "made several curious motions with it that they could not comprehend or myself either." But he maintained his composure, took three hearty whiffs, and passed the pipe to the next person in the circle, at which point "every one gave a great ho three times and the people appeared to be highly pleased at my dexterity with the pipe."

Thirteen years later, on the cusp of the Continental Divide, Meriwether Lewis also received a lesson in smoking customs when he sat down with a group of Lemhi River Shoshones, all of whom removed their moccasins before partaking of the first pipe presented by the American. The practice symbolized a gesture of goodwill that Lewis interpreted as an offer to go barefoot if their sincerity was in doubt. Having frequently complained about the barbed spines of the local prickly pear cactus, the captain considered this "a pretty heavy penalty if they are to march through the plains of their country."

After their initial smoke, the party moved to an encampment just west of the Divide near modern Tendoy, Idaho. There the captain and his three companions were ushered into a small lodge and seated on green boughs and antelope skins. One of the Shoshones pulled grass from a circular area in the center of the lodge and lit a fire. Their headman, Cameahwait, produced a small needle-stemmed pipe and filled the bowl with some of his own tobacco. After a brief delay while the Americans were reminded to remove their footwear, Cameahwait "lit his pipe at the fire kindled in the little magic circle . . . and uttered a speech of several minutes in length." He offered his pipe to the four cardinal points, made as if to hand it to Lewis, then pulled it back and repeated the ceremony before passing it to the captain for his ritual puffs. "On these occasions," Lewis commented, "points of etiquet are quite as much attended to by the Indians as among scivilized nations."

As an amateur botanist who had grown up on a farm that

raised commercial tobacco, Lewis regarded Cameahwait's herb with great interest. He wrote that the Shoshone plant appeared to be identical to a sample he had collected the previous fall from an Arikara garden on the Missouri River. He learned that the Lemhi River band did not cultivate the plant themselves, but obtained it from the Crow Indians east of the Rockies and from other Shoshone bands who lived farther south. A careful drawing of Cameahwait's needle pipe in Lewis's journal depicts a beautiful, compact piece of functional sculpture fashioned from green stone found on the east side of the Continental Divide. The Shoshone chief gifted the pipe to Lewis, along with his own name, which he said meant "Come & Smoke."

After parting from Cameahwait and struggling through the Bitterroots, the Americans met a band of Flathead people. According to Flathead oral tradition, the tribe welcomed the visitors by offering some of their own tobacco, but as soon as the newcomers tried to smoke, they pronounced the Indian tobacco "no good." Lewis and Clark then cut some of their Virginia tobacco and gave it to the Flatheads, "but it was too much for them, who had never tried the American weed, and all began to cough, with great delight to the party." Finally the captains called for some kinnikinnick—a trailing perennial whose small leathery leaves were smoked by many tribes—to mix with their trade tobacco; this the Flatheads found acceptable.

As the Corps made its way down the Columbia, William Clark noted that among the people living downstream from the mouth of the Snake "smoking is by no means so habitual tho' still used in their councils." Beyond the great trading center of the Dalles, however, the practice became much more common, and around the mouth of the river the herb was in great demand. Clark reckoned that blue trade beads were the gold standard there, and white beads were as valuable as silver, while tobacco held a middle rank between them. "The tobacco most admired," he added, "is our tobacco."

In mid-December 1805, Clark exchanged a few fishing hooks and a small sack of Cameahwait's tobacco with some Clatsop men for wapato roots and an otter skin. At the turn of the new year, a group of Skillutes from upriver swapped him some more roots and a neat little rush bag that contained three pipes of their own tobacco. After sampling the new blend, Clark remarked that it was very similar to that of the Shoshones. Less than two weeks later, upon meeting with a Cathlamet band, the smoke blew both ways: Lewis traded some cordage and a plug of his twist for some "indian tobacco" and another basket of wapato. Lewis also recorded the same curious method of inhaling observed by William Broughton. The smokers would swallow several long drafts, holding the smoke in their lungs "until they become suffused with this vapour." Lewis perceived this progressive inhaling, which induced spasms of belching and farting among its adherents, to be much more intoxicating than the usual method of smoking.

Little Plantations ~

CLOSE ON THE HEELS OF LEWIS AND CLARK, Canadian explorer David Thompson crossed the Rockies and began his extensive survey of the Columbia and its upper tributaries. In the spring of 1808, canoeing south on the Kootenai River near the forty-ninth parallel, he peered through his telescope at beautiful meadows and grass-covered hills. "This is the place where the Indians speak so much of growing their tobacco; and we named them on that account, 'The Tobacco Meadows.' " Three days later, he picked up two small turtle shells near the plantings and wondered if they had been used for dipping water. A rich body of tribal lore confirms the importance of these meadows, now known as Tobacco Plains.

Over the next three years, as Thompson established a circle of trade that covered the Inland Northwest, he passed by the Tobacco

Plains several times with no further mention of either the herb or its cultivation. Then in July 1811, two days after launching a canoe from Kettle Falls for his run to the Pacific, he stopped at a Salish encampment near the mouth of the Nespelem River. During the course of a rainy day and a long portage, Thompson shared five feet of trade tobacco with the people there before pleading that his stock was depleted. The Nespelems told him that "they now intend to pull up a little of their own tobacco for smoking, though not yet ripe." But Thompson was in a hurry, and without waiting to sample the local weed, he stepped back into his canoe to paddle away.

For the remainder of his trip to the mouth of the Columbia, Thompson made a point of stopping to smoke with every band he passed, carefully noting the details of their various ceremonies. Among the Methow people he observed that the women were allowed to join in the smoking, though they "were permitted no more than a single whiff of the Calumet, whilst the men took from 3 to 6 whiffs." Near the mouth of the Wenatchee River, "a very respectable old man sat down by me thankful to see us and smoke our tobacco before he died." At Priest Rapids, as a line of dancing women steadily advanced on the furmen, "all the time the men smoked, & like the rest something of a religious nature." When he reached the mouth of the Snake River and was shown peace medals that Lewis and Clark had handed out on their earlier journey, Thompson cut off two feet of trade tobacco for a headman, and calculated that since the previous evening he had smoked and given away a full two and a quarter fathoms (almost fourteen feet).

When North West Company clerk Alexander Ross stopped at the great summer gathering at the Dalles fishery in 1812, he saw people from the interior bartering homegrown tobacco along with horses and buffalo robes for dentalia shells and other items from the coast. Around the same time, a clerk at Astoria heard about tribal tobacco grown both up the Willamette drainage and at the Tobacco Plains on the Kootenai River.

In 1819, after spending a season in the Snake River country, trader Donald McKenzie reported that some of the Shoshone bands he met there preferred their own tobacco to the company brand. In trade, the Shoshones also held their herb in high esteem: While willingly exchanging a beaver pelt for a small pocket mirror worth twopence, they demanded an ax worth four or five shillings for a pint of their tobacco. McKenzie described the wild tobacco as a short plant of a brownish color that sprouted in the sandy, barren soil of their country. To prepare the leaves for smoking, the Shoshones rubbed them between their hands or pounded them with stones until they reached the consistency of green tea. Even though weaker in strength, he considered the native version a good substitute for manufactured tobacco, with the same "aromatic flavour and narcotic effect," although it did leave a certain green taste in his mouth.

McKenzie also delved into the lore surrounding tobacco among the Snake River Shoshones:

> There is a fabulous story current among these people, and universally believed, that they were the first smokers of tobacco on the earth! That they have been in the habit of using it from one generation to another since the world began. That all other Indians learned to smoke and had their tobacco first from them. That the white people's tobacco is only good for the whites: and that if they would give the preference to the white people's tobacco and give up smoking their own it would then cease to grow on their lands and instead, a deleterious weed would grow up in its place and poison them all.

Shortly after David Douglas arrived at Fort Vancouver in 1825, he learned that some of the tribes on the lower Columbia were cultivating tobacco, and he began a quest to track it down. Douglas thought he had succeeded when he saw a native at the Dalles with a tobacco plant in his hand, but the man refused to give up his prize for scientific inspection; even when the botanist offered him a plug of manufactured tobacco in exchange, "he would on no consideration part with it."

A couple of months later, while botanizing along the lower Willamette River, Douglas stumbled onto a "little plantation" where the herb was growing. Without hesitation, he clipped some specimens and headed back to his camp. On the way he met the owner, who recognized the source of Douglas's sample and was visibly displeased. But when Douglas offered two finger-lengths of his trade tobacco, the man became more friendly and was soon talking about his plants. His people did not cultivate their tobacco near camps or lodges, he said, "lest it should be taken for use before maturity. An open place in the wood is chosen where there is dead wood, which they burn, and sow the seed in the ashes. . . . He told me that wood ashes made it grow very large." Douglas was very taken with the idea of using wood ashes to feed the soil, and he offered his informant another finger-length of manufactured tobacco. "When we smoked," the naturalist concluded, "we were all in all."

On a later collecting trip near the mouth of the Columbia, Douglas's Chinook guide brought along a personal stash of tobacco. With each succeeding rest stop along their journey, Douglas grew increasingly amazed at the amount and intensity of the man's smoking. When the botanist was forward enough to tease him about it, the guide offered him a pipe. "In self-defense I was obliged to smoke, when I found that my mode of using the Indian weed diverted my companion as much as his had me. 'Oh,' cried he, 'why do you throw away the food? (smoke). See, I take it in my belly.'" Douglas, coughing helplessly, sucked in a food grown by a people who tended no other crops.

Taken together, these early accounts provide a rough picture of tribal tobacco cultivation across the Columbia country, from the Kootenai River to the Snake and from the mid-Columbia to the mouth. It is also clear that there was an active trade in the herb back and forth across the Rockies and up and down the Columbia. But little is known about the identity of the plants being cultivated.

The tattered leaves and withered flowers of the plant that David Douglas pilfered from the garden on the Willamette are the only preserved specimens of tobacco gathered from the Columbia drainage anywhere near the time of contact. Botanists have identified his pressed plant as a variety of *Nicotiana quadrivalvis*, commonly known as Indian tobacco. A large-flowered, vigorous shrub with long, tubular, white blossoms, Indian tobacco is not native to the Northwest; its natural habitat is centered along the Pacific slopes of California. Through trade and cultivation, native peoples extended its range far to the north and east. Along its journey, the ancestral species developed many varieties, with slight differences in appearance or taste that were clearly recognized by the tribes. The specimens gathered by Archibald Menzies on the Queen Charlottes in 1787, Meriwether Lewis from the Arikara village in 1804, and David Douglas on the lower Columbia in 1826—as well as samples taken from Crow tobacco gardens in Montana later in the nineteenth century—are all currently recognized as cultivars of *Nicotiana quadrivalvis*.

From the samples of Indian tobacco gathered along the Columbia and Missouri Rivers, it seems logical to conclude that this was the herb being used throughout the Columbia drainage. But there is another species of tobacco in the intermountain West, and this one is indigenous to the region. Coyote tobacco (*Nicotiana attenuata*) is a smaller-leafed, scrubbier plant than Indian tobacco;

the tubes and petals on its cream-colored flowers are noticeably shorter. The natural range of coyote tobacco spreads north from the Mexican highlands through the Great Basin, then gradually diminishes as it laps over the Continental Divide and up through the Columbia Plateau.

How widely coyote tobacco was used by people in the Columbia drainage remains unclear. It might be the same herb that figures in a Wishram myth in which Coyote creates people from the scrapings of his tobacco pipe. Ethnographic accounts gathered from Sahaptin and Interior Salish peoples describe all manner of medicinal treatments and ceremonial uses, varying from band to band. Coyote tobacco was used to soothe itchy scalps, cuts and sores, eczema and hives; mixed with bear fat, it made a balm for burns. Chewed leaves relieved toothaches, breathed smoke opened up clogged nasal passages, and a weak suffusion helped expel worms. A strong suffusion of the plant's green leaves made a powerful emetic.

Some groups told early anthropologists that they harvested coyote tobacco opportunistically, while others recalled traditional gathering sites visited from year to year. Coyote tobacco, tough enough to survive in the volcanic scablands of the interior, would appear to be the plant that Donald McKenzie described growing spontaneously in the Snake River country. A few ethnographers have speculated that coyote tobacco was the species being grown on the Nespelem River and at Tobacco Plains. Others disagree, citing the extensive trade networks that would have distributed the larger, more smokable leaves of Indian tobacco. Given the information available, it is impossible to say whether any particular tribe was growing Indian tobacco, coyote tobacco, or both. It is hard to draw a line between a plant that grows spontaneously and one that is consciously gardened; between a knowledge of what the land produces and the forethought of cultivation and harvest.

The Much-loved Weed ~

LIEUTENANT WILLIAM BROUGHTON, remarking on the fondness of Chinooks for tobacco in 1792, suggested that "it might become a valuable article of traffic amongst them." Subsequent sea captains obviously shared his opinion, for when Lewis and Clark wintered at Fort Clatsop in 1805, they reported tobacco as one of the items that Indians on the lower river were obtaining from visiting ships. When fur posts were established along the length of the river, manufactured tobacco became one of their most important imports. The ceremony of smoking continued to be a necessary prelude to conducting business, and traders journeying about the country made sure to stop at any camp they passed for a "friendly calumet," for to omit this tribute was considered an insult according to the customs of the Columbia. Agents unrolled carrots of leaf tobacco as rewards for good behavior and used the promise of tobacco to influence diplomacy between tribes. When competition heated up between rival fur companies, one trader scheduled a "grand smoking match" to lure customers to his post.

In addition to its importance in the social scheme, tobacco also became a significant element in the regional economy. There had been a lively trade in native tobacco long before the arrival of white men, but those transactions had been measured in pipes; now bales of "the much-loved weed" arrived by ship and canoe. Traders purchased salmon and sturgeon, deer and elk, canoes and timber for varying lengths of twist. They paid young men a few handfuls of leaf tobacco to lug supplies and boats around portages. David Douglas considered tobacco "the currency of the country" and used it to hire guides and rent horses. Other naturalists exchanged plugs of tobacco for live birds, young wolves, and a great variety of other specimens.

The example of the white men, many of whom had a perpetual

pipe in their mouths, coupled with the relative abundance of a commodity that had been previously scarce, changed the role of tobacco in native life. Gradually the habit of smoking overlaid the ancestral ritual of smoke. Many came to regard "the envied plant" as a necessity for everyday life in addition to a sacrament for ceremonial occasions. Travelers at the Dalles were greeted by men begging for "pi-pi," and tribes in the interior grew resentful if supplies at the fur posts ran low. When trader Ross Cox arrived at Spokane House in the summer of 1813 after an absence of several months, a large crowd gathered, and the Spokane chief delivered an oration:

> My heart is glad to see you. . . . We were a long time very hungry for tobacco; and some of our young men said you would never come back. They were angry, and said, "The white men made us love tobacco . . . and now we are starving for it."

Touring the Columbia region in 1835, missionary Samuel Parker was impressed with the ways in which the tribes mixed a small amount of tobacco with other herbs so as to conserve their supply, and he admired their dignified manner of sharing a pipe while discussing business or telling stories. But as time went by, he became disheartened at the effects of the "stupefying vegetable" on the culture, and decided that it was far too expensive an indulgence, a vice from which they should be rescued.

⊶ ⊰⊱ ⊷

As manufactured tobacco flooded into the region, traditional garden plots began to fade from view. By 1829, when questioned about tribal horticulture, Hudson's Bay Company agent John Work reported from Fort Colvile: "The only instance of agriculture I have heard of in the district is among the Kootenais where, with great ceremony, a small quantity of a kind of tobacco is sowed." After the time of Work's report, evidence of tribal tobacco cultivation is extremely hard to come by. Systematic botanical exploration around the Northwest in the late nineteenth and early twentieth centuries failed to turn up any Indian tobacco at all. When anthropologist A. F. Chamberlain constructed a list of plants used by the Kootenais in the 1890s, the entry for *Nicotiana* was followed by a conspicuous blank. By 1912 collector William Cusick was convinced that the plant once traded from the Continental Divide to the Pacific was "not to be found in the Columbia River Basin region." Apparently this cultivar was not hardy enough to naturalize itself so far north, and soon after people stopped saving the seed and replanting it in the spring, it faded away. Those tribal people who wished to carry on the tradition did so in plots that they did not show to visitors from outside the tribe.

Even as Indian tobacco disappeared from sight, the indigenous coyote tobacco persisted around the tributaries of the mid-Columbia. Between 1890 and the 1930s, botanists collected specimens from the lower Snake River and the Columbia's Big Bend; along the lines of Wilson and Crab Creeks in the heart of the Basin; from the San Poil and Nespelem Rivers; and up the Okanogan into British Columbia. Some seeds floated far downstream; during World War I, a researcher plucked several sprawling specimens of coyote tobacco from sandbars in the river between Portland and old Fort Vancouver. The era of dam building drowned many of these collecting sites, and modern disturbances have altered much of the habitat where coyote tobacco might have grown. Current botanists maintain that coyote tobacco in the Columbia Plateau has always been an ephemeral plant, with no specific habitat keys and no guarantee that a healthy patch will reappear in the same spot from one year to the next. It is a hard flower to pin down.

━•━ ═◄╪═ ━•━

If you follow the Kootenai River upstream past Tobacco Plains and into British Columbia, it veers east around a bold nose of rock, leaving a short portage to Columbia Lake. There, with formidable mountains rising on either side, the mother river begins in a shallow marsh. It is a quiet place laid upon geological turmoil, a warm pocket surrounded by long winters, a place where wet and dry habitats bump into one another. As I scrambled around the lake's roadless side, I touched both arid rubber rabbitbrush and a delicate cliff brake fern. Junipers crept beneath Douglas firs that looked as if they belonged in a rain forest. Above a metamorphic outcrop, I ate dust along a weedy trail that suddenly gave way to rills of clear liquid hissing from the hillside. It looked like the kind of landscape that might be the birthplace of both a great river and a special plant. Kootenai lore specifies this area as the source of their original tobacco seed, but there were no plants to be found in the places I looked, and the longer I gazed across the vast expanse of mountain and wetland encompassed by the Rocky Mountain Trench, the more I realized I was going to need some help to find them.

At a tribal dinner north of the international border, a friend introduced me to a Kootenai elder as the man who might set me straight. He was a small, wiry figure with perfect posture who displayed a Veterans of Foreign Wars pin in his neatly shaped hat. As I rattled on about David Thompson's tortoiseshell dippers and the two kinds of tobacco I had been researching, his face revealed nothing. When I wondered if he, or anyone he knew, might want to go looking for the plants, he waved me silent with a slight turn of his hat. His voice was extremely quiet; he spoke very slowly, so that I could understand every word. But the meaning of those words floated past me until some time later, when I was tracing the Kootenai River back through its corner of Montana.

I had stopped beside the highway to look at a couple of large rocks that were covered with artwork and caged by steel bars to prevent vandalism. Among the figures were a pair of identical buffalo, executed with amazing precision. One stood right side up, the other upside down, and each animal displayed wonderful curved hooves. On one of the steel bars in front of the buffalo, just above the ground, someone had tied a strip of red cloth. Inside the bars, perched on fresh dirt, lay a broken Camel cigarette. This pair of traditional offerings might be seen at tribal sites across North America, and I wondered what place a wild plant could hope to hold in the face of such machine-rolled abundance. Staring at the brown flakes of tobacco spilling from the torn wrapper, I heard the words of the Kootenai elder as if for the first time.

"What you don't understand yet," he had said. "What you might never understand is that this is more than tobacco we're talking about here. This is us. And I don't believe you'll be able to tear the two of us apart."

When I had opened my mouth to reply, he silenced me with another twist of his hat. "Let it go," he said. "We say let it go."

Alkali ~

WHEN I RETURNED TO THE MID-COLUMBIA, I didn't feel the need to pry into anyone's private rituals; I simply wanted to see if I could find coyote tobacco growing in the wild. By visiting herbariums around the region and poring through early anthropology reports, I made a list of historic collection sites and began visiting them opportunistically, bouncing from a landfill at Boardman to a manicured lawn at the old Almota ferry landing on the Snake, from a corral on Wilson Creek to a poison ivy paradise in the Ginkgo National Monument. I homed in on year-round springs, oases that produced odd bird sightings, and little buttes that looked like pleasant campgrounds. More and more, I noticed that the old coyote tobacco sites coincided with slight wrinkles in topography. Whenever I arrived at the kind of terrain that held promise, I would trudge contentedly for mile after mile without ever seeing anything like the plant I was after.

One August afternoon I found myself walking the reaches of a wide coulee near the western limits of the Columbia Basin. Even though the sun had tipped well past its midday height, the air temperature still hung around a hundred degrees, and I had trouble synchronizing my feet as I cut across windrows in a freshly cut wheat field. "Watch your step," the rancher who had given me permission to explore his land had cautioned. "Bull snakes tend to lay out beside the piles, but rattlers like to crawl up into the straw."

I made for the inviting coolness of a heavy-lidded depression in the entablature of dark basalt. As soon as I stepped into the shade of the uplift, I could see the entrance to a deep cave that had been hidden from sight by a mound of loose dirt. An arch of finely crystallized pillow basalt formed the portal, its top rim decorated with withered cliff brake fern.

Inside, the cave opened into a dramatic chamber. Enough light shone through the entrance for me to see that the ceiling was black, perhaps from smoke, with patches of white showing through here and there. Beyond a shelf of clean basalt, a smaller arch opened into a tunnel that was plugged with dirt and rocks. A mummified deer mouse, large body and long tail intact, lay untouched in the center of the room. News articles about cases of hantavirus had appeared in the papers over the past week, and I wondered if I should tie a hankie over my mouth. But beyond that, I could not quell a rising sense that I did not belong in this cave. I dropped to all fours and scrambled over the berm, brushing the dangling ferns.

Back in the glaring sunlight, I watched a red combine follow a prancing border collie into the field and begin to chew on a last corner section of unharvested wheat. As I approached, the rancher I had spoken with earlier wrenched his machine out of gear and stepped from the elevated cab, happy to tell me what he knew about the cave.

In the 1930s, the landowner had found a bundle of sagebrush matting hidden in a crevice in one of the rock walls. He unrolled the bundle to discover a carved cedar carrying case about a foot long; inside, a beautiful stone calumet, straight with a belled end, lay in a bed of soft twisted cedar bark. The case and the pipe were both inscribed with markings reminiscent of rock art along the nearby Columbia. Some expert—the rancher could not remember exactly who—identified it as a Columbia Salish headman's special pipe, and said elders used the cave to hide tobacco from kids who hadn't earned the smoking privilege.

Geologists who followed the pipe-finder had been thrilled to find evidence of a secondary lava flow inside the cave. The rancher's favorite moment came the afternoon one of the scientists announced that they had identified a sticky kind of tar on the back wall of the cave as pack rat urine. It had drizzled in from an unseen chamber, and provided the clue that led them to a hidden lava tube. "Pack rat piss," he chortled as he swayed with his idling combine. "Can you imagine?"

The border collie jumped up to lick my hand, and I turned to scope the canyon that curled behind the headwall and the cave. When I asked about the watercourse, the rancher pointed toward the skeletal remains of a barn and a century-old walnut tree, planted by his granddad the year he built his original homestead. Rocking to the intricate rhythms and knocks of his machine, he poured his knowledge of the coulee down on me. If I caught the gully of a seasonal creek just below that walnut tree, he said, it would lead me all the way to the top of the ridge. I was welcome to walk up there and look for all the flowers I wanted. In fact, I should keep an eye out for a spring box at the far end of the property—the outlet pipe was always getting plugged, and he would appreciate it if I scraped the weeds off the screen on my way past. That sounded like a deal to me.

"The water up there might look good," he warned me as I departed. "But don't drink too much of it. We don't want to be hauling you out of the canyon on a rack."

The heat had finally begun to ease by the time I started up the draw. The past spring's heavy runoff, its path clearly marked by a white band of alkali, had carved deep channels through the sandbars. Falling into my familiar search pattern, I zigged this way and that, working slowly up the braids. A battered irrigation pipe ran down the high side of the watercourse, established long enough ago that clumps of serviceberry and mock orange had sprouted beside dependable leaks. A yellow-breasted chat, its face masked like an outlaw, chortled at me from the shadows. I laughed back, then stopped to admire a pale evening primrose, still blooming after all the other wildflowers had withered in the sun.

My first rooted-in-the-earth coyote tobacco plant appeared near the top of the alkali band edging one of the sandbars. In a landscape of washed-out weeds and pastel primroses, its greenery flashed like a sunning reptile. Every surface of the plant was covered with a golden fuzz, those infamous glands of nicotine. The entire specimen measured a modest twenty inches tall, but on the hottest day of the summer it still managed to sport sixteen fresh blossoms, each one a creamy trumpet not much more than an inch long. Ten more had recently faded, but the tobacco had not yet set any of its distinctive seed pods.

I had circled the plant twice, fully satisfied with the wonder of finally finding one, when I stumbled upon another. It was puny compared with the first, a single unbranched stem with only a couple of new blossoms to show off. This tobacco sprouted from the creek bed proper, a place that on several occasions during the winter and spring must have run hard with water. But the runt showed me something that its sister plant lacked: a dozen fully ripe seed capsules, some of them already cracked open and shedding their tiny seeds. In my hand they looked like the yellow-brown nits combed out on lice days at elementary school.

Still working my way upstream, I found a total of six plants within fifty yards. The largest one was a healthy bush more than two feet tall with three dozen blooms and two dozen faded flowers. At the base of its many stems, several brown crinkled leaves represented the cured remains of the plant's first burst above ground. Crushed in my hand, they exuded the scent of air-cured tobacco.

Shadows were climbing the canyon wall by the time I returned to the main coulee. With night coming on, the wind began to flow easily down the canyon, repaying the thermal debt of the morning and freshening my dusty mouth. I relocated my first clump of coyote tobacco and sat down to wait for darkness, scratching figures in the sand and thinking about how water coursing from the canyon spring, boosted each year by bursts of spring runoff, must have provided enough moisture to sprout a few seeds in the gravel year after year.

Venus peeked over the east rim of the palisade, chased by the light of a waning moon. I leaned down every few minutes to drag my nose across the small blossoms of the coyote tobacco. Before the moon caught up with Venus, a faint scent of honey seemed to drip from one white trumpet, then dissipate in the night air. I thought I must have imagined it, and dipped my nose to the flower once more. No matter how close I twisted my face, I couldn't catch the smell again.

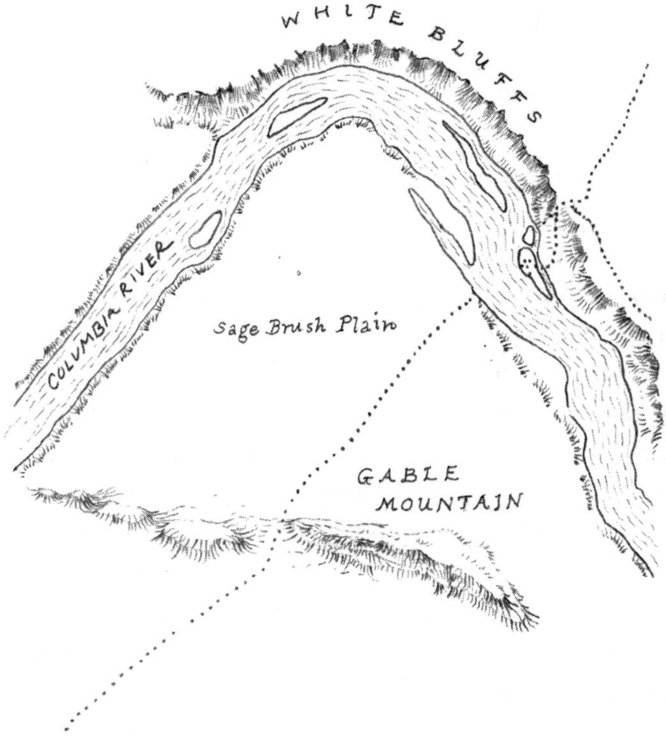
WHITE BLUFFS
COLUMBIA RIVER
Sage Brush Plain
GABLE
MOUNTAIN

CHAPTER NINE

The Otter Swims

The old White Bluffs trail, after Alfred Downing's Map of the Upper Columbia River, 1881

The Vessel~

LATE IN THE twentieth century, the elegant hyperbola of the
Trojan Nuclear Plant's cooling tower rose as a signature landmark
on the lower Columbia River. Steam pouring from its maw mirrored
the concurrent activity of Mount St. Helens, and the plant, located
on the Oregon side just upstream from the mouth of the Cowlitz,
churned out enough kilowatts to light the metropolis of Portland.
But after two decades of service, small faults began to plague its operation—a balky alarm system, a whiff of radioactive gas, and, as the
last straw, cracked steam tubes that shut it down for good in 1993.

Six years later, the plant's dissemination began in earnest. By
then, the reactor vessel had already been stripped of its zirconium
fuel rods, sealed at every orifice, and encased in a jacket of high-
quality concrete. Eleven-foot foam impact limiters rimmed with
stainless steel were attached to each end, then the entire package
was wrapped in heavy blue plastic. All trussed up, the vessel
resembled the stubby femur of some impossibly large mythical
beast, weighing just over a thousand tons and measuring forty-two
feet long by seventeen across. The two million curies of radiation
inside the wrappings caused a stir of anxiety among some residents, although that figure represented only about 1 percent of
the radioactivity in the spent fuel rods left behind on the Trojan
site, and paled in comparison with the 185 million curies released
in the Chernobyl incident. A General Electric spokesperson estimated that a person standing within six feet of the shielded reactor for one hour would catch about the same amount of radiation
that an airline passenger received from the sun on a transconti-
nental flight.

The vessel was carefully loaded on a barge and pushed into
the shipping channel. During the night, tugboat and barge cruised
upstream past the bright lights of Portland, then hissed quietly

through the locks at the Bonneville and The Dalles Dams, passing from the wet Pacific Slope to the sparsely populated lands of the interior. The shipment cleared John Day and McNary, and turned the corner at Wallula Gap. Early on Monday morning it reached the confluence of the Snake and Columbia Rivers and chugged through the Tri-Cities to the Port of Benton, 270 river miles upstream from its starting point beneath the cooling tower. For the next nine hours, a curious crowd on the receiving dock watched as workers unbolted the 20-axle, 320-wheeled transport trailer from the barge, lowered the craft to meet the ramp, and hooked a mammoth tractor to the trailer's hitch. It was early evening when the tractor towed its load up the grade to Horn Rapids Road, leaving considerable ruts in the gravel.

Once on the road, the crew held their speed to a steady pace of five miles per hour as they crossed the Pasco Basin. "You don't want to get up any momentum of any sort," commented the utility spokesperson. It was midnight before the trailer reached its destination near the center of the Hanford Reservation, where the relic from the nuclear age drew to a halt in the shadow of Gable Mountain, another relic from another time.

The Heart of a River ~

ON A HOT SUMMER AFTERNOON, sun rays bounce off the flat bottom of the Pasco Basin, casting a tangible haze. The illusion of a shallow iridescent sea spreads across the expanse, and it is the easiest thing imaginable to float away on the mirage, buoyed by the shimmering waves of heat, back through a couple of hundred millions of years to the edge of an ancient ocean, warm and swarming with life. During that long-ago time the copious remains of clams, snails, tube worms, and trilobites layered the seafloor with thousands of feet of marine detritus. To the east lay the solid land of the mother continent; from the west, groups of islands sailed in to collide with the mainland, riding atop a large tectonic plate that was sliding beneath the edge of North America. Ichthyosaurs swam in the huge bay encircled by these new additions, and pteranodons with ten-foot wingspans swept overhead.

The sea receded to the west, leaving a wide coastal plain. The air was warm and humid, and frequent rains pelted the land; streams and creeks meandered across the gently sloping surface, collecting to form the ancestral Columbia and Snake Rivers. A water-loving rhinoceros wallowed in lakes, while ancestral horses, camels, and elephants found their niches in this Eocene world.

Around twenty million years ago, clouds of steam appeared in the southeast, and a black wave of viscous lava rolled across the landscape, filling river valleys and lapping against the edges of the surrounding highlands, then cooling into a giant plate of dense basalt. The displaced Columbia wound along the northern edge of the flow, creeping through blackwater cypress swamps as it searched out a new path to the sea. Staggered pulses of magma spread new layers of hot lava across the region; again and again, the flows obliterated plants and animals and pushed watercourses back to the edge of the plateau. Time after time the Columbia cut a new course through the rock, and life inched back like a slow tide across the basalt. Eventually the flows abated to occasional trickles; by then the basalt measured thousands of feet deep, and its great weight had depressed the very crust of the earth.

Wrinkles appeared on the floor of the tableland like folds in an unruly cloth pushed from the southwest. The uplifts enclosed an irregular bowl near the western edge of the plateau, the beginnings of the Pasco Basin. Almost lost among the larger folds, a narrow ridge stretched for half a dozen miles across the building floodplain of the Columbia River. The river's braided channels supported forests and open woodlands, providing habitat for a small deer with exotic mooselike antlers, a medium-sized ground sloth, and a panda that resembled its reddish relative that lives today in southeast

Asia. Herds of Pierce's peccaries plowed through the Pliocene vegetation, rooting for nuts. Occasional mastodons rumbled past.
Eruptions of Cascade volcanoes dusted the basin with thick coat-
ings of ash. On the edge of a pond, turtles plodded through the
mud, surrounded by scuttling crabs; frogs croaked from the shal-
lows, while catfish and sunfish flashed past. In the open river, white
sturgeon cruised beside suckers and peamouth chub.

The growing Cascades cast a rain shadow over the basin.
Grasses and shrubs covered the drier areas, providing food for
lantern-jawed camels and the American Zebra, one of the first
members of the horse family to develop that lock in the upper fore-
leg that makes it possible to sleep standing up. Bone-eating dogs
roamed this world the way the hyena courses across the African
veldt, as both predator and scavenger, competing for prey with a
bear and at least two different cats. A weasel family member somewhat larger than our modern otter also worked the scene. Its rear
molars rose and dipped in three sharply peaked crowns that
inspired the species' Latin name, *Trigonictis*. These molars might
have nipped after ground squirrels, rabbits, young beavers, or fish;
such a varied diet would have required an agile carnivore that
could both climb and swim, one whose hunting ground covered a
whole range of habitats.

The weather began to cool, and cold winds swirled off the
lobes of the great ice sheets that slid down from the north. Many of
the animals died off, and the river piled mud and silt over the last
of their bones. Packs of bone-eating dogs remained to dine on
holdover herds of camels along with Pleistocene mammoths, long-
horned bison, and shrub oxen. As the glaciers began their final
retreat, violent floods swept in from the east, transforming the
Pasco Basin into a huge lake not once but several times. Successive
deluges drowned the little ridge in its center until the water worked
its way through Wallula Gap and downstream to the sea. Then the
promontory peeked out across a flat-bottomed lake bed filled with

gravel and sand, embraced by a semicircular curve at the bottom of the Columbia's Big Bend. The small mountain, swept clean, stood at the heart of the great river.

Nuk Say ~

THE LONG RIDGE THAT snakes across the pancake flatness of the Pasco Basin is labeled on maps as Gable Mountain, but the Wanapum people who lived in its shadow called it *Nuk Say*, The Otter. Indeed, when viewed from certain vantage points, a knobby head rises above the basin floor, followed by a rounded back; a sloping rump disappears beneath the surface for a mile, then a small butte mimics a broad tail flopping back into view. It is the shape of a playful paddler stretching up for a look around.

Climbing its rump on a breezy morning, I paused partway up the rocky slope to admire the graceful arc of the Columbia River sweeping across the foreground, touching three points of the compass as it curls south to meet the Yakima and the Snake. This section of river abounds in salmon and sturgeon, and people have netted fish here for at least sixty-five hundred years. I looked down at dark drifts of sand piled against The Otter's flanks by the prevailing winds. Several years ago a pipeline surveyor taking a break on one of these dunes noticed a scattering of bone fragments and fire-cracked quartz. Archaeologists collected other artifacts that indicated a kill site, and dated the materials back to two thousand years before the present. One team conjectured that a group of hunters from a nearby village, carrying mussels from the river to ward off hunger, might have come across a small group of bison and herded them between two large sand dunes near the edge of the ridge. The hunters hurled darts with their atlatls, dispatching eight animals, then butchered them with heavy stone choppers and cooked part of the meat over a fire. When they departed, they left behind several stone tools and a pile of mussel shells.

In the centuries since that bison kill, the wind has continued to batter the ridge, sculpting shrubs and pitting rocks with airborne sand. Lowering my head into the relentless breeze, I reached The Otter's broad back. Fractured escarpments along the spine of the ridge testified to the forces that folded and bent the mother rock back upon itself. The vegetation was low and open, a variety of tidy bunchgrasses offset by the deep olive of stiff sagebrush and the dusky rose bracts of summer hopsage. I passed a series of lichen-spotted cairns, marking out the long stretch of time during which the uplift has been both a spiritual and a practical landmark for a host of mid-Columbia tribes. For many generations, boys and girls of various bands have come to Nuk Say for spirit quests, and there are certain places on lower talus slopes where rocks have been carefully piled to create sheltered blinds, perfect spots to sit and watch for game.

The biting wind drove me back toward the swimmer's midsection, where two lines of rock diverge to form an open cavity lined with grass. I settled into the protected nook and looked across the Columbia at the long curve of the White Bluffs rising steeply above the river. In the millennia since the last of the Pleistocene floods swept downstream, the river has steadily sliced a cross-section through its old floodplain, opening to view many pages of the basin's history. As the river eats away at the bank, it occasionally reveals the antler of a false elk or the tooth of some long-extinct horse. Near the beginning of the bluffs lies a large island where modern horses, first acquired by the Wanapums through trade with tribes to the south, grazed in the shadow of their ancient ancestors.

The Spanish horses presaged the arrival of other newcomers. On a summer day in 1811, a cedar plank canoe flew through the rapids just upstream from The Otter, bringing the furman David Thompson, who stopped long enough to share a pipe and promise a trade house before sailing on downstream. But his description of the arid steppe and a note in his journal—"of course there can be no beaver"—made it clear that the desert held little real interest for furmen.

Aside from the fur brigades that paddled through twice a year, pausing only to camp and purchase fish for dinner, the Pasco Basin remained relatively quiet for another four decades. The pace of change suddenly quickened in 1858 with news of a gold strike on the upper Fraser River. Hundreds of men swarmed up the Columbia in canoes and skiffs; others came on foot or horseback, following a well-established Indian route from the Yakima River. Cattle drovers hooted thousands of cattle along the trail between The Otter's rump and its flailing tail, then swam them across the river and headed north to British Columbia to feed the hungry miners. Campfires of sagebrush and greasewood dotted the flats at night. A Nez Perce chief rode to the White Bluffs with word of a skirmish on the prairies to the south, and Yakama warriors painted their faces in preparation for battle, but the Wanapum kept the peace, ferrying footsore miners across the river in dugout canoes in exchange for a handkerchief or a shirt or a fifty-cent piece.

The next year the sternwheeler *Colonel Wright*, named for the military man who was busy subduing the Plateau tribes, churned past to unload mining tools and provisions at the foot of Priest Rapids. An entrepreneur set up a ferry at the White Bluffs, powered by long sweeps; mule teams pulling freight wagons dashed past The Otter, then waited in line for three days or more for the ferry to transport them across the river to the dusty road leading north. Another steamer came on line, its giant paddle wheel pausing at a new dock beneath the White Bluffs while bags of gold dust were transferred from stagecoaches for the trip downriver. A storehouse, blacksmith shop, and saloon joined the ferry owner's driftwood cabin on the wide beach beneath the bluffs.

As long as the goldfields held out, the sounds of wranglers and livestock filled the basin. The tribes began to complain of rising dust clouds and erosion, and exotic weeds like cheat grass and tumble

mustard began to crop up on overgrazed terrain. In the village beside
Coyote Rapids, just below Nuk Say's tail, the Wanapum prophet
Smohalla plaited strips of otter fur into his hair and performed the
Dreamer dance. Smohalla beseeched his people to reclaim their
spiritual heritage in the face of encroaching white civilization. If
they could only hold true to the land, he said, the world would turn
over and all that they had lost would come alive once again.

And then, as if his dream were coming true, the pack trains
slowed and the cattle drives stopped. Save for a few Chinese miners
who filtered in to pan for gold on the bars below Coyote Rapids,
the crush of traffic ceased. Smohalla hoisted an American flag
above his lodge and turned down the repeated entreaties of government agents to move his band to the reservation. A cattleman
named Hank Gable began pasturing horses on one of the river's
grassy islands and running stock around the base of The Otter.
Because it was the only prominent feature in the flat basin, early
settlers identified the landmark with the rancher who worked it,
and government surveyors inscribed the names of Gable Butte and
Gable Mountain on their maps. One of them declared of the area,
"a more dismal place it would be hard to imagine."

But some people were taken with the stark beauty of the setting, and through the 1890s a few stalwart settlers trickled in,
drawn by the long growing season and rich soil. Farmers took up
land along the river between the ferry landing and the large
Wanapum village at Coyote Rapids. The newcomers dug wells by
hand and cleared sagebrush to plant melons, fruit trees, vegetables,
and sugarcane. "Anything will grow if we can only get water," they
said. An eastern promoter enlisted investors and had begun digging
a ditch to irrigate the land between the Rattlesnake Mountains and
Gable Mountain when the financial panic of 1893 evaporated his
capital. There was too much water in the spring of 1894, when a
heavy runoff swept houses down the river along with livestock, furniture, and wagons. Those who stayed opened a school on the

point of land below Gable Mountain's shoulder and dug pieces of clay from the bluffs to use as chalk. A schoolteacher arrived, sunbonnet tight around her head, and watched her students assemble by horseback and rowboat. The aged Smohalla, blind and ill, mounted a horse and was led by his wife along the old trail beneath Nuk Say on his final trip to the Yakima Valley. Hank Gable purchased a new ferry upriver, floated it through Priest Rapids at high water, then hitched some of his horses to its newfangled treadmill.

Even though the Pasco Basin receives only six inches of rain a year, the dependable flow of the Columbia and a reputation for growing good fruit eventually began to lure homesteaders and speculators alike just after the turn of the century. A few miles beyond The Otter's nose, the geometric lots of a new townsite were platted by Judge Hanford of Seattle, and dirt began flying from the ditches of his ambitious irrigation project. In the years between 1907 and 1909, promotional brochures advertised the area as the "California of the Northwest," and steamers were crammed with settlers headed for White Bluffs and Hanford. More sagebrush was cleared, fields plowed, and orchards planted; more dust storms swirled across the basin.

In 1913 the Milwaukee & St. Paul Railroad pounded spikes into a spur line, inspiring the town of White Bluffs to move four miles west to meet the rails. For the next three decades the two small towns of Hanford and White Bluffs competed and socialized. Fourth of July parties were held on the ferry, and watermelons were cooled in the river in summer. During cold winters, children skated on the ice; in one particularly frigid year they were joined by two thousand sheep, a fraction of the twenty thousand that wintered between Gable Mountain and the river. Sahaptin youth still climbed Nuk Say for vision quests, and picnicking homesteaders scrambled to its summit to look down on their blossoming orchards and check the progress of a large substation being built just upstream to receive power lines from Grand Coulee Dam.

Caravan ~

ON A CLEAR December day in 1942, a small plane bearing a colonel of the U.S. Army Corps of Engineers flew in from the west, crossed the Pasco Basin, slowly circled the valley twice, then disappeared back over the Rattlesnake Hills. Two months later, an explosion of activity erupted around the post offices of White Bluffs and Hanford. Shocked residents held registered letters announcing that the United States government would be purchasing their property and over six hundred square miles of the surrounding countryside. They had thirty days to vacate their homes and farms. The purpose was top secret, a war effort originally called the Gable Project, later changed to the Hanford Engineering Works.

Within weeks, engineers carrying stakes crisscrossed the basin as people packed their belongings. Coffins were disinterred from cemeteries; trucks hauled furniture and livestock away. The Wanapum fishing villages near the White Bluffs were closed. Rows of barracks and Quonset huts quickly overran the Hanford townsite. Buses arrived daily from Pasco, bringing construction workers from all over the country. An access road was cut up the face of The Otter so that a radio tower could be installed on top; heavy transmission lines were strung across its back, and a hole was blasted in its side. Railroad tracks and dusty roads spiderwebbed the flats below. Between the landform and the river, enormous buildings began to rise from pits excavated into the deep sand and gravel left behind by the glacial floods. Fifty thousand workers swarmed around the buildings.

Early in 1945, the massive buildings were completed and the population tide at the construction camp turned: Loaded buses pulled out, and barracks were dismantled. Soon sand blew across the patios of the deserted trailer park as goats wandered the empty streets of Hanford Camp. Scientists continued to arrive, traveling under assumed names, and security grew tighter. Every so often during the spring and early summer, a caravan of three cars would emerge from one of the large buildings and proceed directly to the side of Gable Mountain. Two army officers carrying a small steel container would climb from the center car, approach a heavy steel door fitted into a short concrete wall, dial the combinations of two locks, and enter a vault carved into the base of the mountain. Minutes later, they would exit, swing the great door shut, and spin the locks. At regular intervals the process would be reversed, and a container would be removed from the vault and loaded back into the caravan.

On August 6, 1945, the *Richland Villager* issued a special edition headlined "It's Atomic Bombs: President Truman Releases Secret of Hanford Product." The vault beneath Gable Mountain, it turned out, had served as the temporary storage chamber for the enriched plutonium used to fuel the A-bomb that was built at Los Alamos, New Mexico, and dropped on Nagasaki, Japan.

For the next four decades, throughout the grinding tensions of the Cold War, the Hanford area continued to live a secret life as new reactors were built along the curve of the river. By the late 1980s, however, all of the units had been shut down, and the focus shifted to nuclear power research complexes and decontamination laboratories. The Gable Mountain vault stored soil samples rather than plutonium. As attention increasingly turned to the accumulating deposits of radioactive material around the country, Hanford was named one of three locations to be considered as a national long-term disposal site. The Basalt Waste Isolation Project, or BWIP, envisioned repositories in stable basalt 2,500 feet below the surface, where spent fuel rods could be safely stored forever. Results from test wells drilled in the Pasco Basin during the Cold War held great promise—wildcatters had bored through 7,600 feet of basalt just east of the Columbia and struck crystalline basement rock; near Rattlesnake Mountain they had stretched the limits of their drill rigs through 10,660 feet of basalt without ever breaking through.

From the start of the BWIP project, geologists knew that many variables would have to be considered, and The Otter slithered back into the picture. A pair of hundred-foot adits were drilled into its western flank, then packed with geologic and hydrologic sensors. One of their major purposes was to learn how the heat given off by nuclear waste materials might affect the structure of the existing basalt formations near the river.

The Otter was not entirely cooperative. When snowmelt from the ridge began dripping onto the equipment in the adits, researchers realized that surface cracks penetrated surprisingly far into the bowels of the mountain. Deeper rocks turned out to be under more pressure than expected. Sensors recorded recent movement in the supposedly stable basalt. One fault had been generated by an earthquake estimated at 5.5 on the Richter scale that had occurred within the last two thousand years. To geologists with nuclear disposal on their minds, that was far too recent. In 1988, the stability of the rock became a moot point when the U.S. Congress selected Yucca Mountain, Nevada, as the approved disposal site. Authorities shut down the experiment on Gable Mountain and bulldozed the rubble into an approximation of its original form.

That shoulder of the mountain remains abandoned today, and the old plutonium storage vault is empty of everything, according to one Department of Energy official, except snakes and lizards. Within a couple of hundred yards up the steep slope, yellow and orange lichens begin to crawl again, and ancient stacked cairns seem to mark a defiant boundary: the big machines worked to this line and no farther. From here on up, The Otter prevails. Any stopping point along its ragged ridge still provides a breathtaking panoramic view of surrounding landmarks, old and new. The undulating profiles of Rattlesnake Mountain, Umtanum Ridge, Wahluke Slope, the Saddle Mountains, and the Horse Heaven Hills circle the horizon. To the north, neat rows of weatherbeaten locusts and Chinese elms outline the former streets of White Bluffs. Off the north side of Nuk Say, a skein of sand blows across the dunes where bison were butchered two thousand years ago. All around Hank Gable's old grazing grounds lie the square concrete hulks of mothballed nuclear reactors. The White Bluffs, their mudstones and volcanic tuffs changing in color and texture throughout the course of the day, outline the river's course. Eating steadily away at their layered past, one of the last free-flowing stretches of the Columbia, designated as the Hanford Reach National Monument, sweeps past. In the center of it all sits The Otter, still seeking protected status on the National Register in recognition of its cultural importance to native peoples and its pristine shrub-steppe communities.

⊶ ≡◆≡ ⊷

The Hanford site is accustomed to handling the hot remains of a long cold war, and rectangular scrapes in the basin floor mark the burial sites of many decommissioned nuclear parts. In the summer of 1999, large machines clanked onto a special disposal area south of Gable Mountain and began digging a trench 850 feet long, 150 feet wide, and 45 feet deep to receive the reactor vessel of the Trojan Nuclear Plant. On the morning of August 19, with the hole finally finished, the tractor steered the trailer bearing that vessel to what was hoped would be its final resting place. Beside the massive ditch, U.S. Representative Doc Hastings gave a brief speech, then waved a small flag, signaling a drag-line operator to drop three yards of backfill onto the stubby hulk. Many of the people gathered for the ceremony were acutely aware that they stood on the brink of a pit whose effectiveness would be measured over epochs. Some spoke hopefully of the advances in disposal technology that might take place in a decade, or a century, or a millennium. But it was a scale of time that only The Otter, swimming above them all, could really comprehend.

CHAPTER TEN

A Good Day for
Digging Roots

Canby's lovage, or licorice root *(Ligusticum canbyi)*

Crickets ~

"CAN YOU SAY THAT AGAIN, AUNT ALICE?"

Alice Ignace moved her lips, and the unmistakable chirp of a cricket emerged. She made the sound with no apparent effort, barely smiling when the class dissolved into shivers and grins.

"That's our word for cricket," she said, repeating the call. "You try it."

Several members of the class took a crack at the pronunciation, trying to roll s and r sounds across their tongues in imitation of Alice. Some of the tribal kids added clicks and stops, and a couple managed a fair imitation of the sequence. But none of them could evoke the music of natural sound and spoken language the way Alice had.

"That's good," she said, laughing with the kids at the cacophony. "But you have to keep going. It takes a long time to learn how to speak Indian, and I need some more people to talk to. Come on, get your mouths working. Like this." She stuck out her chin, and the whole class leaned forward to read the complex mechanics of lip, tongue, and teeth meshing together. The cricket sang again, and they all tried to chirrup back.

Alice Ignace is an elder of the Kalispel tribe who grew up speaking Kalispel Salish. Her grandparents told stories that hearkened back to a time when that language constituted the core of a different world. Although the Kalispel Reservation now consists of a strip of land only a few miles long bordering the lower Pend Oreille River in Washington state, Alice feels at home up and down the entire river, from its curl through British Columbia, to Pend Oreille and Priest Lakes in the Idaho Panhandle, to the Clark Fork and Flathead drainages of western Montana. That was the country her people used to traverse, back and forth, according to their seasonal needs and desires. When I began visiting the public school across the river from the Kalispel Reservation to talk about the history of the drainage, Alice would sometimes come by and show us how large a watershed could be.

On this day, after Alice finished with her cricket demonstration, we turned to the subject of the Canadian explorer David Thompson and his first contact with a Kalispel band in 1809. When I read Thompson's journal account of the meeting, Alice pointed out that he had neglected to mention that the Kalispels refused to trade with him until he sat down and ate some of their dried salmon and camas cakes, a story told to her by her grandmother. Over the ensuing winter, Thompson had compiled a list of Salish words and phrases in his journal, practical terms like "dried meat" and "do you know who took my horse?" When I wrote some of his broken phonetic syllables on the chalkboard so everyone could try to pronounce them, Alice shook her head at the dullness of his ear.

I recounted how the explorer spent time sitting with Kalispel elders in their winter camps in Montana, and how during that winter he wrote short phrases in his journals such as "Bears hum when they lick their feet."

Several kids in the class immediately raised their hands—what did the humming bears mean? Here Alice agreed with Thompson for the first time. "He's right," she said. "Haven't you ever been out picking huckleberries and seen the way bears do it? Maybe your David Thompson should have tried talking to the bears."

She stuffed her fingers into her mouth and fluttered them up and down against her tongue and lips, making the most delightful *numanumanumanuma* sounds. The class all crammed their fingers into their mouths and began humming along with her.

As we talked about the arrival of English, French, Eastern Woodland, and even Hawaiian people into the Northwest, Alice explained how the Kalispel language bent and flowed around the host of new languages brought by the newcomers. Cree men from across the mountains were called "scouts," because they often preceded the wave of British traders. Iroquois trappers were given a name taken from their distinctive haircuts, which at first sighting made women and children draw back in fear. Farm animals like roosters and pigs entered the language as approximations of their French forms. As the place changed, the language adapted as well, but there was always a core that remained Kalispel.

Little Fingers ~

ALICE WAS OUT FEEDING HER DOGS when I pulled into the driveway, but she already had on her digging outfit: jeans, plaid flannel shirt, and neck scarf, with a wide white eyeshade topping it all. On the clear fall morning a heavy frost lingered on the ground, and leaves were tumbling off the cottonwood trees. Across the Pend Oreille River someone was splitting wood, each knock echoing across the water like the report of a fired gun.

"Good day for digging roots," Alice offered, smiling as she stepped to the front door to grab a pair of work gloves and her short iron digging stick. From the stoop, she introduced me to a pair of new puppies in a box under the eaves, warning me not to get too close because their orange dingo-chow mother stalked a nervous half-circle twenty feet away.

"She's wild," Alice said. "She's always been wild." She coughed deeply, then tapped her chest with one palm. "I'm getting over a cold. Maybe I'd better get some of my medicine."

She opened the screen door and went inside the house to grab a tin of Drum tobacco; back outside, she pried open the lid. Resting on top of the rough-cut Drum was a bracelet strung from knuckle-sized pieces of a dried root. Alice tore a knuckle from the string, adding a sharp tang of celery to the mellow aroma of yellow-brown tobacco; she popped the morsel of root into her mouth and nibbled it with her front teeth as she slid the bracelet onto her wrist.

"Now I'm ready," she announced, clutching her lunch sack under her arm.

We headed north, toward one of her favorite bald mountaintops, a drive that began quietly with half an hour's ride along the river. As we rounded a curve, Alice pointed to a spot beside the road where she had found a little porcupine curled up a few weeks back, not much bigger than a tennis ball, with its quills still so soft that she scooped the thing up and brought it home. Her friend Bernice asked her what she was going to do when its quills got sharp: How would she shoo it out of the way? So Alice returned the porcupine to the spot where she had found it, and the mother was still there, guarding three other little balls. Alice had set the fourth one down on the ground and talked to the mother until it said *ssshew ssshew*, taking her little one back. The porcupine language took me back to our recent day in the classroom, when the range of Alice's voice made it seem possible to communicate with anything, and it reminded me that I had brought something to show her.

On the seat between us lay a plastic bag that contained a cluster of roots from a plant I had found south of Spokane. Alice picked up the bag and squeezed its contents briefly. "Oh yeah," she said. "Wild carrots."

"Yampah?" I asked.

"That's what some tribes call them, but it's not our word. To us they are Little Fingers, see?" She pronounced the Kalispel word, *s-luk'wm*, as she pulled one of the roots from the bag and held it up so I could see how the shrunken, elongated tubers twined around one another just like little fingers. "I haven't seen any of these around here for a long time."

As we came around a curve that revealed a long view of the river below, Alice pointed back upstream. "You know before that dam raised the river, there was an island right out there," she said. "Every year in early summer, that's when my grandmother would call me to go with her. We'd row over to the island in her little boat and find plenty of those wild carrots. She'd keep me working until we had almost a gunnysack full. They were so much easier to dig than camas. When she decided we had enough, she'd tie up her sack with a cord, then tie the cord to the boat line. The sack would drag under the water, see, and the river would wash those roots all the way across."

Here Alice used her hands to show the action of the water, one hand closing against the open palm of the other, back and forth, back and forth, washing the soil away from the s-tluk'wm. "By the time we got to our side of the river, they would be all cleaned off. Then she would open the bag up right there and we'd eat wild carrots—didn't roast them or anything, just ate them as they were. After we'd had as much as we wanted, we would spread the rest out on a mat to dry. In a few days they would shrivel up, all skinny and twisted, almost as small as these that you have here. When she wanted to eat some over the winter, she'd drop them in a jar of water and you could watch them blow back up until it was like they had just come out of the ground. She never cooked them even then."

Alice was silent for a while. "I miss those little fingers," she said. "They were her favorites, you know."

Alice never named her grandmother, even though she talked about her frequently. They had spent a great deal of time together when Alice was a child. Sometimes her parents and siblings would go down to Newport and spend the night, and since Alice didn't care much for going to town, she would stay home with her grandmother, who insisted on speaking Kalispel the whole time.

"My mother told me always to eat whatever food my grandmother fixed me," Alice said. "She would make cakes of camas and deer meat and different berries and different kinds of fat. My dad liked bear fat the best. She always put a little surprise in—serviceberries, huckleberries, wild carrot roots, a little sugar. You never knew what they would taste like. And she would sit and do things, always busy. People would bring her a deer hide or a beaver or an ermine for their dance outfit, and ask her to fix them out nice. She would use brains to tan them, but she used other ways too. She'd stretch a little white ermine and sit there all morning, rubbing it with a cake of bear fat that fit in the palm of her hand, not smoking or anything, no TV, just rubbing that cake back and forth to make the ermine's skin soft."

By now we were passing through a section of the national forest that had been recently logged. The fallers had left many young tamaracks on wide centers, each one ringed with blue paint. They looked like spindly teenagers who didn't want to get too close to one another. "Sure looks different without all the trees," Alice said. But the heavy cut did afford us a clear look uphill at our destination. The bald peak rose blocky and round to the northeast, and we could see the last lines of trees threading toward its top before petering out altogether.

"People ask me why I go all the way up there to get my roots, and I smile because my grandmother told me that the best roots grow up there, the most powerful of all around. On top there's the juniper, which she used for everything. She'd boil up the branches, twigs, needles, berries, all together and drink that tea if she was getting sick. Or she'd put a bunch of it in her little tub and soak her feet when the arthritis was hurting them. It feels so good."

Alice had come up this same road when she was a little girl— first on horseback with her grandmother, then with horses pulling a small wagon. Her aunt had an old flatbed truck, but they still took the horse and cart because it did better on the rocks. Alice said that if the kids got restless, her grandmother would tell them stories. She used to tell one about a cave up in these hills that was well hidden, with an opening that looked out over the whole river valley. Kootenai raiders who rode down from the north to steal Kalispel horses would hide out in the cave, waiting for night to come. When everything was silent, that's when they would come down. Alice laughed at how that story used to spook her—those

Kootenais, with their completely different language, were like bogeymen for Kalispel kids. And yet she knew plenty of Kootenai people, saw them all the time at powwows.

"Yeah. My grandmother knew everything about all those old days, had names she called all the rocks and special places. She always told us to be careful which ones we moved and what we dug up, because there might be something there we didn't know anything about. That's why when these archaeologists come around and want to dig everything up, we don't always agree."

We were bumping along through a brushy clearcut now, but Alice cocked her head back down toward the river valley.

"We had a lady down here, she died during that epidemic of scarlet fever or typhoid, whichever it was. After she was gone they buried her right beside the house and burned it, her whole house, with all her things inside. There's an archaeologist who has worked with us before, he's a good man. He called to ask if he might have permission to dig that lady up because he heard she had been buried with lots of artifacts that might tell us all kinds of things about our culture then. We elders sat and talked about it all one evening. After a while we decided to say no, we didn't think so. Let's not dig her up."

Alice retreated to silence as we moved through the clearcuts. I took the gentle, steady sideways shakes of her head to mean that she was nodding off, and I tried to drive slowly. Suddenly her arm flew up and pointed to a small snake sunning itself in the middle of the dirt track. It was a garter snake, and I stopped and caught it for Alice's inspection.

"Ahhh," she nodded, up and down this time, and the word for "garter snake" slid out of her mouth like the reptile itself: "*s-chew'=ile?*"

The little snake twisted around my forearm, a youngster out for one last jaunt before the cold weather sent it underground. Alice watched it move across the ditch and up the hill.

Jesuit missionaries who arrived in the Kalispel country in the 1830s were the first to try to make written sense of the Kalispel language in a systematic way. Gregorio Mengarini published a primer of Salish grammar in 1861, and J. Giorda brought out his *Dictionary of Kalispel or Flat Head* in 1877–79. Both of these works attempted to shoehorn the tribal language into Latin forms, a difficult task because it has no verb "to be" or any concept of plural nouns. Modern linguists who try to interpret the Jesuit notation can only guess at many points of pronunciation.

When Norwegian linguist Hans Vogt visited the Kalispel Reservation for a few months in 1937, he reaffirmed what speakers like Alice have long known: Flathead, Pend Oreille, Kalispel, Chewelah, and Spokane are closely related. Linguists refer to these dialects as members of the Interior Salish language group, spoken by people along a curving stretch of territory known as the Salish Arc. Contact with European and other tribal languages has eroded the subtle shades of linguistic structure along this arc, and elders complained to Vogt in 1937 that none of the young people could speak good Kalispel anymore.

A contemporary linguist, attempting to construct a more comprehensive dictionary than Vogt had time to compile, has worked with elders on all the reservations along the Spokane, Pend Oreille, and Flathead Rivers. After a few hours of conversation with Alice Ignace, he was impressed with how little her pronunciation has been colored by English sounds. Her Salish vowels still ring clear, as do the deep-throated *w*'s and *y*'s that provide a foundation for the rhythm and pace of the language. He noticed that she was using her pharynx and larynx in ways that very few tribal speakers do anymore. All those days spent listening to her grandmother had infused Alice with an unusually pure strain of the Kalispel language. But the magic of language sparks only when it is spoken, and Alice, like the elders of a previous generation, laments the limited opportunities she has to use her grandmother's tongue. Of the 250 Kalispels on the current tribal roll, only a handful can converse with her in the old way.

⊷ ≍◆≍ ⊷

When we reached the crest of the mountain ridge, I eased to a stop. The moment we clambered out, road-stiff, Alice spotted a shiny-leafed bush. She made me bury my nose in it—nothing smelled so fresh, she explained, as red-stem ceanothus. Her grandmother would boil it, leaves and roots and all, and use it to rinse her beautiful thick hair until it shone as bright as the leaves.

While we gathered a plastic grocery bag full of the ceanothus, Alice was assessing the bear grass that spotted the slopes around us. She slipped on some brown cotton gloves and showed me how to twist up a braid of stems and tear them out, being careful not to slice my hands on the backward-pointing barbs. She turned the bunch over to show me the green-to-white-to-purplish colors at the base of each stem. "This was my grandmother's favorite material for weaving baskets," Alice said. "When she wanted to add some color into her baskets, she would pick a bunch of Oregon grape berries and rub them all over her bear grass."

In the days when Alice and her grandmother had traveled by horseback along this rocky trail, they had taken their time, and Alice saw no reason to up the pace now. She leaned down to sample some of the shiny black huckleberries that grew low to the ground, her favorite kind. She pointed out ground squirrel holes and looked for rabbits. We watched copper butterflies work in the open sun. We let ourselves be swallowed up again in the trees, nice-sized spruce and tamaracks dripping with the black coyote hair lichen that Kalispel people roasted in the earth with camas and skunk cabbage.

"She was kind of bossy, I guess," Alice said, circling back to the subject of her grandmother. "If I was sick she would boil up the juniper berries and twigs and needles and make me drink that tea. Then she would bundle me up in the bed and say, 'Sweat it out!'—rough like that. She had her own canoe that she paddled around everywhere, and if she caught any kids hanging around it she'd grab a stick and come out swinging it around, scare those kids away quick. Every day she would make somebody row across the river over to Cusick to get her an apple pie. There was a little drugstore over there that sold pies, and she had to have one every day. Finally I learned how to make apple pies myself and after that she wouldn't leave me alone."

Bears in the Spring ~

THE TEMPERATURE WAS STILL COOL when we arrived at a lush, shady spring that dribbled steeply down the hillside. This was the place where Alice's grandmother had taught her to dig the roots of the lovage contained in her medicine bracelet. We collected digging stick, gloves, and another grocery bag and slipped down the weeping banks. I splashed through a morass of water hemlock, ferns, columbine, bog orchids, and tough purple monkshood blooming beneath a fallen log, but found none of the lovage's droopy white umbrella flowers among them. Alice wasn't doing much better.

"They're hiding," she chuckled from above. "All I have is this one little baby." She straightened the bent stem of a small lovage, no more than eighteen inches tall. Without ceremony, she worked the point of her iron stick into the ground and lifted, then lifted again, raising a mound of black peat strung with root tendrils and small gravels. Her gloved fingers dove into the mound, twisting and reaching, to snap off the root, releasing a smell of celery into the air.

"Aw, yeah," said Alice. "This'll be a good one."

Now I could see better. Just past the monkshood log, on a circular flat, I spotted two waist-high plants with several stems. On her way down to where I stood, Alice slipped. Straddling a spruce root, one foot sinking into the mud, she propped herself up with her digging stick and laughed. When she got herself picked up, she methodically extracted some roots that were finger-sized and larger, real celery. Together with a mass of muddy threads, she tossed them all into the plastic bag.

"That's all roots," she said. "Know what I do with the little ones? Mix them in with my tobacco and then when I roll a cigarette I smoke them together. That's good."

Alice has a way of saying *good* with low, honeyed o's that made me want to take up smoking. But she certainly wasn't greedy about her roots; when I searched around and found several more clumps, she waved me off.

"No, no," she said. "I got enough in here to get me through the winter. We'll leave some for the bears—they like to eat them too, you know. These and mountain ash berries, that's what they eat to plug themselves up before they go to sleep."

On the way back up the hill we passed an inviting pool of water, and there Alice plopped down on a fallen hemlock to light her first cigarette of the day. When I ribbed her about her cough, she swung the small bag of lovage roots toward my face.

"I'll be all right," she said. "Long as I keep some of this around."

As Alice puffed, she probed beneath a mossy overhang with her digging stick, reaching out toward a small clump of feathers floating in the pool. To our amazement, the clump separated into five tiny bodies, little winter wrens on the verge of fledging. Their scraggly feathers were not fully grown out, and we saw no apparent cause for their death. We looked at them for several minutes, while Alice finished her cigarette.

"I wonder if those little birds were sick," she puzzled. "Some kind of bad sickness." She considered that idea. "The reason we didn't let the archaeologist dig up the house that burned down is that lots of those bad diseases stay in the dirt. That's why people used to burn anyone who died of disease, because some sickness can be carried through the air and infect others. And even when they are buried in the ground, no matter how long they stay down there, the disease is still there with them and you can stir it up. It's still dangerous."

Alice poked at one of the fledglings. "My father, he used to talk about this epidemic all the time, that happened when he was still a boy. Typhoid or scarlet fever, I can never remember which one it was. But it happened around 1918 or '19, before I was born.

"He said the people were dying something awful, three or four every day. The children especially, so many children died. My dad was a young man then, and for some reason he didn't catch it. Blind Paul and Paul Isaac, neither of them got it either. They were elders, so they went around and did what they could. My dad was younger, he helped them out. It was so hard. They'd go into a house and find the people just lying there, they couldn't even get up to eat. In the morning they would find two people dead and put them in their wagon and drive them up to that cemetery between Usk and Cusick. They would dig a hole and bury them, and then sometimes by the time they got back there would be two more people dead and they'd have to turn around and drive right back up to the cemetery. Once they walked into a house and a family of seven people was living there, and all seven of them were dead. *Paauh.* And they couldn't do anything but dig a big hole and dump all seven of them in there, no coffins or nothing.

"It shook up my father, you know, having to bury all those people. He said they dug so many graves in that cemetery that the people were buried so shallow you could probably see their bones sticking up out of the ground. That always worried him, thinking about that." Alice shook her head.

"My grandmother, for some reason she didn't get sick, her and four other women. They did what they could, making soup and other things and carrying it around to the sick people. But most people wouldn't eat. She couldn't even get anybody to go up to the mountains to get any roots or junipers to make good medicine with. No one was strong enough to go up there. So they just did what they could, going from house to house, and more people were dying every day.

"After everyone that was going to die had died, and everyone else was all healed up, my dad said one doctor did show up. No one had ever seen him before; no one knew where he came from. He started handing out medicine like cod liver oil, castor oil, aspirin, stuff like that. He was the only doctor the people saw. My dad always talked about that doctor, and how he and Blind Paul and Paul Isaac had to drive those bodies two at a time."

━━◆━━

The aftermath of that sickness was certainly a low point for the Kalispel people. When linguist Hans Vogt worked with the Kalispels in 1937, he extensively interviewed a man he called One-eyed Tom, whom Vogt guessed to be about eighty-five years old. At one point Tom told him, "The Kalispel are very few. Their roads are covered with grass and bushes. We saw a man walking on the road, crying."

But these were a people determined to survive, and in the decades since, many things have changed in a positive way. The Kalispels established a tribal development program to foster economic security; a tribal singing group has carried Kalispel songs to the far reaches of the continent; earnings from a casino business have been directed toward education. In the last few years, plans have been drafted for an interpretive center that will celebrate the heritage of the tribe.

Alice is interested in the interpretive center as a way to keep the culture going. They have a piece of land that looks out over a wide wetland to the Pend Oreille River. She heard they might place one of their sturgeon-nosed canoes, covered with white pine bark and sewn with chokecherry lashing, in the middle of a display hall. All around it would be arranged precious old things: sacks and baskets and beads and doeskin dresses of the kind her grandmother wore. There would be a banner hanging from the ceiling that said, in Salish: "We Are the People of the River, and We Are Still Here." Off to the side there would be an archive of oral and written sources, and rooms where language and crafts could be taught. Maybe the words could keep them all together.

Alice got up from the pool of dead wrens and walked steadily back to the van. Soon the road broke out of the woods onto an open slope that laid the whole Pend Oreille drainage before us, with the river slipping by as a silver ribbon far below. At the edge of the bald mountaintop and the line of spruce and lodgepole pine, Alice pointed to the exact spot where her family used to camp.

"Oh yeah. We'd set up right here, and walk over to that little seep for water. There was a better spring in the woods down the far slope, on the Priest Lake side—a nice big pool that was full of water, with cottonwood and aspen trees around it. But my dad would go down there and say, 'Shoot, that water's all dirty cause the bears have been in it.' Those bears would roll and thrash and make it so muddy us kids would have to walk across this way, to the seep. Right along that trail right there."

A clear trail through the grass was still visible, leading to a clump of vine maple and ninebark whose leaves were already starting to turn.

"The big spring was about half a mile down the hill. My dad, he got tired of those bears messing it up so he made a little spout out of two pieces of wood and stuck it in the side of the hill right where the smaller spring bubbled out. Then we had good clear water.

"We'd camp here for weeks at a time. Dad would hunt deer and grouse and we'd pick berries. We'd dry some of them, and my mother and grandmother would put their huckleberries in quart glass jars. One year my grandmother tore her big wool blanket into little pieces wrapping all those jars for the trip back down the mountain."

Alice wanted to get on to the top and see her junipers, so we struggled upward, completely out in the open now, scaly rocks running up under the tires, open grassland above and below. Fat grasshoppers lurched back and forth on the dusty tracks, with migrant water pipits bobbing after them. The road bent through an extensive stand of subalpine fir before making one final switchback that led to the summit.

The vista dragged us in opposite directions—while Alice peered west toward her riverside home, I boulder-hopped across the east face, stepping in and out of crystallized snow and checking out all the prickly clumps of mountain juniper. Each one crawled low across the rocks, ten or twenty or thirty feet in diameter, and not one in a dozen showed the first hint of a berry. I knew Alice liked to have some berries. Far below, the surface of Priest Lake shone black and gray like cold steel, and beyond the lake, Idaho's taller peaks carried a light blanket of new snow. To the north, the big mountains up in British Columbia looked as though they were already in the grip of a hard winter.

I circled back to find Alice settled into a thick patch of juniper that was absolutely loaded with smoky blue berries. Her body hovered over the clump, then bent to methodically snap off sprigs and stuff them into her bag. I offered a pair of garden snippers, but she waved them away and continued to soundlessly twist and tear at the tough branches. Once the bag was full, she was ready to go, just like that. While I took one more stroll through the rarefied air, Alice sat in the front seat and caught her breath.

A small flock of birds whistled past my head, twittering away. One of them landed a few feet from Alice's open door. It was a pine grosbeak, rosy-headed and thick in the body. *Chee-fli*, it called. *Chee-fli*, over and over.

When I walked over to it, the grosbeak did not fly.

"Hey Alice," I called, "There's somebody out here calling you. *Al-ice Al-ice*."

"I hear it," she said, and chirped out another Salish word. "That means 'sneaks up on you.' My grandfather used to tell me that bird will sneak up on you and start calling. Some Indians don't like that, the way it just comes out and starts talking. Some others think it's a good sign. It was good for my dad. We'd see them all the time when we used to come up here."

We left before the grosbeaks did, and when we reached the open rocks, we paused to trace again the line of Alice's trail from her former huckleberry camp to the seeping spring.

"That's the way we used to have to walk," she said, "whenever the bears muddied up our good water hole. And my grandmother, she would come out in the morning and whistle until the horses would come up so we could take a ride. I guess we always rode up here as long as she was alive. After she passed on, that's when we stopped. She had her special whistle to call the horses, and after she was gone we could never get them to come when we wanted them. Nobody knew how to talk to them just right."

CHAPTER ELEVEN

Musquash

Tracks of muskrat feet and tail

A Winter Journey ~

THE WINTER WAS AN UNUSUALLY HARD ONE in the Pend Oreille Valley, with snow piled so deep that families could not find enough to eat. All over the valley, members of the Kalispel tribe were suffering. One man decided that before he grew any weaker he would try to cross the mountains to the west and look for food. His wife packed up the last of their root cakes, which gave him just enough strength to make it over the crest of the mountains. He was barely conscious when he descended into the swampy Colville Valley. So goes a story told to Salish elder Antoine Andrews by his great-grandfather.

When the Kalispel man reached the small river that drained the valley, he discovered that the bottoms were locked in a tight freeze. All through the frozen marshes, muskrat lodges glistened silver, and the "push-ups" around their breathing holes rose like clear domes glazed with ice. As Andrews described how the weary traveler surveyed the muskrat village, his story bubbled with a Salish term of amazement sometimes translated as "Wow!"

The Kalispel man found a place to build a fire, and knowing that his everyday bone arrowheads were not strong enough to penetrate the tough hide of a muskrat, he hafted special flint points to his arrows. With his weapons prepared, the hunter carried a heavy stick to the closest push-up and shattered the icy dome that covered the muskrat's breathing hole. Then he crouched down and waited.

A muskrat approached the breathing hole, curious as all muskrats are curious, and stuck its head up to see what was going on. The man shot it with an arrow. He brought the rodent over to his camp, skinned it, and skewered it on a stick of red-osier dogwood. He drove the stick into the ground close to his fire, where the meat could slowly roast. When its fat began to splatter down onto the coals, he returned to the push-up and crouched down again.

Another muskrat peeked out of the hole and soon was skewered next to the first one. The man kept hunting until he had a whole ring of muskrats staked around his fire.

As the meat finished roasting, he ate his way around the circle, grease dripping from his mouth and hands. Wow! That was good food for a starving man. When he was finished, he picked up his stick and walked to the next push-up and smashed the ice cap that covered it. Before long he had another ring of muskrats skewered around his fire. He ate all of those, too, and kept on smashing and roasting until he couldn't eat any more.

Over the next few days, he shot and skinned and dried as many muskrats as he could carry, then took them back over the mountain trail to his home. He divided the meat among all the families in his village, and invited anyone who wished to journey with him to this swampy, food-rich place. Thus a group of Kalispel people came to live in the center of the Colville Valley in the early 1800s, drawn there by little muskrat, the fruit of the wetland.

<hr />

Viewed from the rim of quartzite cliffs that line the west side of the valley above the town of Chewelah, the Colville River has a primordial look. Depressions marking ancient oxbows wind through a maze of backwater pools; wet deciduous growth loops gracefully around a craggy hill that was clearly once an island in a wetter world. Not so many millennia ago, the southern lobe of the last great glacial advance lay in the valley like an extended digit pointing due south. When the ice withdrew, it left tightly curved rills of cold water from one side to the other, a moist fingerprint that formed the perfect habitat for small channel swimmers.

Muskrats are rooted in such swampy North American habitats; they have been adapting to the whims of frozen seasons from the very beginning of their time on earth, back at the far edge of the Pleistocene. These water rodents live all over North America, and their pointed noses and curious habits appear in tribal stories from Hudson Bay to the mouth of the Columbia. Often Beaver and Otter are characters as well, but it is always the smaller Musk Rat, Beaver's Little Brother, who keeps diving until he succeeds in bringing up the handful of mud from which all solid land was created. That's Musk Rat, hero in a watery world. In one tale, he is rewarded with plenty of roots to eat, plus a wife to give him a great many children so that he might become more numerous than any of the other animals. But Musk Rat can also be a jester: Another story describes him distracting an archer from shooting his friend Beaver by pointedly rubbing the scent glands beneath his tail that give him his name. That's Musk Rat, scratching his ass and playing the clown.

The Attraction of Muskrats ~

A MID-OCTOBER BREEZE SWEPT THROUGH the cattails as we slid the boat into the Colville River and assumed our places in the bow and stern. At the highway bridge we ducked low, fended off the concrete pillars, and emerged into a solid stand of reed canary-grass backed by rolling pastureland. A cow that had wandered down for a drink chewed her cud and watched us float past without a backward step.

This part of the valley looked very different than the wide marshy bottoms that greeted the Kalispel traveler two centuries before. As white settlers moved into the area in the late 1800s, there was a widespread movement to drain wetlands and create more land for agriculture. Around 1910, a steam-powered dragline channelized this section of the river into straight lines and long sweeping bends. Except for a few weeks during spring runoff, when a thin sheet of placid water sometimes stretches from one side of the valley to the other, anyone coming down from the mountain today would see wide fields of deep black peat bordering the thin crease of the river. And yet, one element of that former world has persisted. Only a few paces beyond the cow's muddy tracks, Steve Schalock raised his paddle to point out a clear swath cut through the aquatic vegetation.

"When it starts to get chilly in the fall, that's what I always think about," he said. "Football and muskrats."

Although Schalock retains a halfback's feel for open space, fur pelts were part of his family's fall routine for much longer than pigskins. His grandfather started trapping weasels in the wheat country west of Spokane just after the turn of the century; he progressed to coyote, then badger, then drifted toward the watercourses to work muskrat and finally mink. He moved to the Colville Valley in 1929, just in time for the Depression, and there his traplines helped him squeak through the hard times.

Like many trappers, the elder Schalock kept records of catches and sales. In the winter of 1934–35, he sold the skins of six badgers, three minks, fourteen weasels, and seventy-three muskrats. He traded mostly with Sears, Roebuck but occasionally dealt with a merchant whose store dated back to the Hudson's Bay Company era. Those mid-Depression rats fetched an average of fifty cents per pelt—significant money in those days. Fur prices often run against the grain of common logic, and when muskrats more than doubled in value the following fall, Grandpa Schalock knew exactly where to concentrate his sets. His account book for the 1935–36 season tallied 134 rats, which sold for $150.90, dwarfing the minks, badgers, weasels, and single skunk listed below them on the chart.

As pelt prices cycled through the next five decades, two more generations of Schalocks absorbed the craft. Steve was still in elementary school when he made his first muskrat sets inside the Chewelah city limits, and his younger brother and sister in turn also learned how to skin and stretch during the intense November season.

"Our grandpa sawed out a bunch of planks in the shape of ironing boards," Steve recalled, "and you would pair them together. By sliding the two halves opposite each other, you could stretch out any sized animal with a perfect fit. Dad always liked to finish the mink himself, because he wanted them to be just so. Mink was kind of his specialty. But he was willing to leave some of the muskrats to us kids, and I got pretty good at them."

The family laid lines north and south of town, checked them assiduously for two weeks, then pulled them so there would be animals left to overwinter. When Steve began socking away money for college, the Schalocks extended their traplines up a few favorable creeks and began to go after beaver as well. For several seasons the family's catch ranged from twenty to sixty minks, forty to eighty beavers, and two hundred to three hundred muskrats. It was cold, wet, physical work, fueled by a sense of skilled camaraderie.

"After a few days, it became kind of a blur," Steve remembered. "Going out when it was still black dark, parking at every bridge, one guy taking off downstream, the other up. I'd always run between the sets, and if there was a beaver, I'd tie it to a piece of rope I kept around my waist and just drag it behind me. Hard to run with two or three frozen beavers weighing you down. And when you went into the water, man, it was hard to warm back up.

"I remember some Saturdays we'd get back midmorning and be in a complete stupor. Sit around, eat, maybe turn on the TV, lots of skinning to do, and all day we'd still be glowing just from being out there."

When the price of rats leaped past five dollars apiece in the late 1970s, other people jumped into the business, but not all of the newcomers were thinking in the long term. It took only a few traps left in the water too long to hurt the beaver population and absolutely ruin the country for mink.

"Muskrats, though, you can't really bother them," Steve said, "because they reproduce so fast."

From the bow of the canoe, he pointed out more open pathways, each one exactly the width of a swimming muskrat. The intrusive reed canarygrass marched along both sides of the river, choking out all native vegetation. Naturally, that was what the muskrats were eating. They had pulled up thick bunches of canary-grass by the roots and arranged them in haphazard piles that served as feeding platforms. By the time we had passed a few of the clumps, grouped like miniature loose haystacks on sunny bends along the shoreline, Steve was beginning to make an estimate of the fall rat population. Some of the ragged masses had broken free to drift downstream, and we caught up to one raft after another. Each mat showed sharply clipped green tops and succulent white root tips, tinged pink where the stem emerged from the ground. Occasionally the licorice-green hair of water milfoil, unsightly with slime, coiled in with the crisp grasses. It was another intrusive alien plant, another unsightly nuisance.

"Yeah," Steve said. "They'll eat that junk too. Crayfish, frogs, dead fish—they aren't too particular."

Over the summer the animals worked the river, caching silage and roots in tunnels burrowed into the bank. When it turned cold, many moved into mounded winter lodges in the broader wetlands, where they could stay inside during bad weather, coming out to use their feeding platforms when it was warm and fair. These feeding stations were situated for escape from predators, yet exposed enough to afford a pleasant nibbling place in low afternoon sun.

I remembered the first time I really watched a muskrat, perched on a shelf like that, devouring green shoots that dripped with the delicious moisture of a muddy pond. The animal manipulated the food so nimbly that to me, as a child, its paws looked like miniature hands and fingers. I couldn't believe it when my Uncle John reached for his rifle the moment he saw the animal, and cursed when it dove before he could shoot. There were too many of the rats around, he muttered, and they kept undermining the dams on his fishponds.

Steve agreed with my uncle. If there was nothing to control the rats, he said, they would get overcrowded and start fighting with each other. You would see bite marks and pus pockets spoiling the pelts, and diseases like tularemia could come in. Sometimes when a pestilence struck you would find whole clumps of muskrats dead in their lodges, and sometimes surviving animals ate the dead ones. Studies had shown that the germs could live in the ground like distemper around a barn, so that after an outbreak you might not see any muskrats back in an affected area for years and years.

"Grandpa Schalock always said that they actually do better if someone's in there trapping on them," Steve said. "It's like they need some pruning."

He steered toward the bank, where a low saddle of mud dropped into a hidden backwater. The rise was slick from muskrat sliding, and Steve was certain there was a tunnel nearby. A rat will burrow into a bank, just beneath the surface of the water, then tunnel upward to excavate a large, dry chamber. Over the course of a summer, a pair will raise as many as four litters of kits in their secure nursery. In the easy-digging black peat beside the river, a tunnel might extend for yards, and after a few years of use, a bank can be riddled with numerous entrances, passageways, and chambers. Cow hooves and tractor wheels can break through from above, and the Swiss cheese banks are much more susceptible to erosion. That was why the Schalocks rarely had trouble getting permission from farmers to trap on their land.

We came around a bend to a dilapidated farm bridge. As we eased beneath the slumping timbers, Steve checked the darkened wedge of shore for sign. A muskrat had been at work in one corner, its distinct long-toed tracks neatly split by the unbroken line of its dragging tail. The prints led directly to a pile of freshwater clam shells. On a bed of rocks the mollusks glowed like gray mussels, empty of any meat.

"Rats like those clams, too. And you'd be amazed at how many of them there are in the river."

As we floated back into the sunlight, a basking muskrat dove off its feeding platform, the long curl of its tail turning over with a slurpy splash. *Musquash*—the Iroquois-Algonquian word that radiated from the Great Lakes along with the fur trade—still sounds about right.

After a minute or two the animal reappeared, motoring along as if it were pushing a paddleboard. The animal's progress seemed measured, almost leisurely, as smooth humps undulated along its back to its rump. A thin wake trailed behind its stiletto tail, the perfect rudder for navigating in slow water. Watching such a lumpy character cruise blatantly in the open on a sunny afternoon made it impossible not to wonder how a trapper might fare along this stretch of river come November.

"Right there," Steve said, placing his paddle across the gunwales to frame a spot on the bank. "That'd be a good place to make a set."

He kept his eyes on the place as we drifted past. "No matter how long it's been since you quit laying traps, you never stop looking."

Traders ~

PACIFIC STEEL AND RECYCLING occupies a sprawling warehouse surrounded by dirt alleyways in the industrial part of Spokane. The smells of hydraulic oil and stale beer mix in an odd, almost pleasant fashion as sellers unload their recyclables on various conveyor belts and watch them roll toward complex smashers and shredders. One winter day a few years ago, I wandered into a back section of the warehouse where bales of green hides—cow, elk, and deer—waited to be shipped. Tightly baled, they constituted forklift versions of the bundles that moved through the doors of the company when it was founded a hundred years ago as Pacific Hide and Fur.

At the top of a wooden staircase above the warehouse office, Roger Scheurer answered his door after a single knock. Dressed in a white lab coat, he moved in quick darts, like a winter ermine: bright-eyed, aware, focused on the long circle of his steady hunt but ready to be distracted.

"Ah," Roger said. "It'll be just a minute. I'm with a customer."

He ushered me inside and returned to his black telephone. The windowless room was arranged like a dry cleaning shop, with parallel rows of overhead racks from which hung hundreds of beautifully finished pelts, and the smell within was feral but soft. Coyotes glimmered in variations of gray, white, silver, rust, chestnut, and dusky yellow. A single dark Canadian wolf dwarfed them all. Bobcats were running high that year, and Roger had a line of them on special display. Next to the cats he had draped a few red foxes, punctuated by a pair on the end so black I wondered if they might be dyed. Four striped skunks and a single wide badger began the transition from long-haired land dwellers to the slender swimmers with their short, dense coats. Beaver pelts, finished in the round, were stacked like throw rugs on the showroom floor, and minks hung neatly off one side of the grading table. Among a dozen river otters, yellow and disproportionately long, one was set apart, and I caught a whiff of unscraped fat. As if he sniffed it across the room, Roger looked up and put his hand over the receiver.

"Don't know why I took that one," he apologized. "Doing the guy a favor, I guess. He's just starting out."

When Roger joined Pacific Hide and Fur in the early 1970s, the company's trade in fine pelts had been on the decline for decades. The Spokane headquarters was shifting into scrap iron and recyclables to stay afloat and considering a name change to reflect their modern ways. Roger started out downstairs with the beer cans, but soon noticed that once in a while some old-timer would drop by, looking for a cash offer on a few furs. Roger purchased a few pelts on his own and found out where to resell them. Gradually, he learned about grading, and finishing, and worldwide markets where different species reigned as favorites. After a while, the company decided to support Roger and, riding the crest of a few high-market years, furs reassumed a small niche in Pacific Steel and Recycling's annual report.

Roger had overseen the construction of his fur room so that it comfortably met his needs. At the moment, a large sanitary table in the center of the room held a stack of cased muskrat hides. They were turned inside out, dried and flattened into the shape of big red bedroom slippers. Roger's assistant, also clad in a white coat, methodically picked up each skin and placed it next to a measuring line on the table. With only occasional rejections, he sorted a dozen fair-sized ones into a pile, tied them neatly with a length of leather cord, and set them down on the floor among dozens of identical bundles. The effect of Roger's room was to collapse time, back to the very beginnings of the fur business and beyond; the warmth and beauty of those pelts blanketed the entire span of human history, with the skin of each species surrounded by its own aura.

When David Thompson crossed the Continental Divide to establish the first commercial trading post in the Columbia District in 1807, beaver pelts were his primary interest, with Beaver's Little Brother relegated so far down the list of desired furs that it barely registered. Of the twenty fur packs (weighing about ninety pounds each) that Thompson collected that first winter, he recorded only "two measly muskrats" within their contents. But he carried those rats all the way back to Lake Superior, and he kept his eyes open for more. By the spring of 1812, as he walked an ancient trail through rich wetlands in the Idaho panhandle, Thompson noted "Many Rats in the Marshes."

A clerk from the Spokane House trading post of 1822 did a brisk business in muskrats, often swapping a few skins brought in by tribesmen for small amounts of tobacco and medicine. On the Hudson's Bay Company's 1824 table of exchange rates for the Columbia District, the muskrat held the lowest value of any item on the list. Rated at one-tenth the value of a prime beaver pelt, it took 180 muskrat hides to purchase a gun, and ten just to buy a dozen fish hooks. But while muskrats were not valuable, they were abundant; their hides might be small, but they were very compara-ble to beaver in quality. In 1824, almost seventy-five hundred rat pelts were shipped out of the Colvile District, and in some years dur-ing the 1830s and 1840s, almost double that number were collected.

⊷ ⊰⊹⊱ ⊷

Roger hung up the phone and introduced me to his assistant, Rolf, who had learned the trade in eastern Europe. I happened to be teaching some kids who had recently immigrated from Russia, and I told Rolf about a red fox cap that one of the boys had brought to school; his uncle had made it from scratch, with warm ear flaps and a button on top.

"Yah," said Rolf enthusiastically. "Those people know their furs. And their foxes, such fine animals to see running through the snow. Not like these devils here." He held up one of the pelts he was sizing and gave it a playful shake.

Muskrats, I knew, were not native to Europe. In 1905 a Bohemian count brought three pairs of muskrats home from a visit to Alaska, and according to legend the descendants of those six ani-mals populated all of northern Europe. In reality, fur farmers on the continent and in Great Britain imported muskrats for several decades in the early 1900s, and escapees quickly established wild populations.

"If you could see what they have done to us," Rolf said. "They are everywhere. They burrow, and too many burrows ruin the dikes. The canals go haywire. When we try to trap them or shoot them, it seems even more come back. Many people at home, they hate your muskrat."

Considering some of the detritus that has drifted across the

Atlantic from the Old World to the New—smallpox and measles, starlings and Norway rats, Russian thistle and water milfoil—it's almost reassuring to realize that the lowly musquash has managed to travel the opposite direction, and it is no surprise that the Old World has not had much better luck controlling muskrats than we have had with milfoil. The only place in Europe where muskrats have been exterminated after introduction is the British Isles, and that was accomplished only by dedicated trapping programs coupled with limited habitat. Places like the Netherlands, with miles of earth-banked canals, have become muskrat havens. In several northern regions, across Scandinavia and Siberia, rat trapping has become a lucrative pastime for many families.

"Ah, yes," Rolf admitted. "They do have a nice pelt." He took one of the cased skins he was grading and pulled it over his left hand like a glove, then neatly turned it inside out so that the hair was back on the outside. Now he had a muskrat puppet, with holes where the feet and eyes had been. The long brown guard hairs bristled along its sides, running to black along its back and light gray under its belly.

"Style?" asked Rolf. "You want style? These little rats look good."

"No domestic market for them, of course," Roger broke in pragmatically. "But we've learned how to adjust. I have to leave myself a little room, you might say. At the beginning of the season I have a feeling for the price of muskrats, maybe three dollars on the high side—that means the real high-quality skins. Then I get a sense of how the season's going: how the weather is, what kind of shape the animals are in, where the demand is coming from."

The price ratio of rats to beavers still runs about one to ten; three dollars for a cased muskrat versus thirty for a blanket beaver. Roger dug out a tally sheet of fur prices from 1973 to the present, and showed me that during the winter of 1979–80, rats were fetching an astronomical $6.32. Exactly ten years later, a prime muskrat pelt was worth ninety cents.

"You take last year," Roger continued. "We paid about three dollars for muskrats, and exported 100 percent of them, the best part of them to Russia, then Poland and Czechoslovakia, places like that. They use them mostly for trim—cuffs and collars on cloth coats, that kind of thing. Now, in Russia they go for hats. I understand muskrat hats are real popular back there right now."

Rolf ran a thumb backward against the lay of the fur. Beneath his touch, the long coarse guard hairs, colored in shades of earthy brown, parted to reveal dense dove-gray underhairs; on a live animal, the interplay between the two textures trapped air bubbles, boosting both insulation and buoyancy. As a pelt, the texture of the undercoat deepened to a bluish, leaden cast more like a woven carpet of goose down than single follicles.

"That's the felt," he said. "Good enough to keep anybody warm. That's why people where there's real winter are wearing one of these on their heads."

Columbia River tribes certainly would have agreed with Rolf. Coastal peoples like the Chinooks fashioned muskrat capes that were waterproof against the interminable winter rains. Several Interior Salish tribes such as the Spokanes and Coeur d'Alenes sewed pelts together with sinew and bark thread; sometimes they cut the skins into strips and wove them on simple looms into robes or blankets that were light but warm. When Rolf ran his thumb along the edge of his puppet, touching the notch where the tail had been neatly removed, he was tracing a seam that joined Salish weavers and Czech furriers. He moved to the opposite end of the pelt, where a patch of nose not much larger than a shiny black peppercorn remained intact. Roger Scheurer, with an eye trained for detail, leaned forward to examine the work around the muskrat's muzzle.

"That's a trapper," he said respectfully. "I like it when they care enough to cut around the nose."

Holes ~

BY THE END OF FEBRUARY, winter should have been relaxing its
hold on the Colville country. But this year, after a messy January
thaw, the weather had turned cold again, and ice glazed every
wrinkle in the valley floor. Skeins of crusty snow hung along the
river's twisted oxbows. The water that had flooded over the wetland
during the thaw had frozen cleanly, leaving large open sheets that
looked like perfect places for some boot skating.

Close up, however, I found jagged rifts and shattered edges
that were anything but smooth. The freeze-up had dropped the
water a foot or more, cracking the perimeter of the marsh into
long shards of glass. This was the frozen kingdom that had greeted
the Kalispel traveler after he crossed the eastern mountains. As I
started across it, I spotted a familiar shape tucked into a line of cattails—the mound of a muskrat's winter lodge, ragged with yellow
cattail stalks and clumps of snow. When I walked toward it, the ice
began to crack and sing beneath the balls of my feet. One particularly loud line shot clear across the marsh, changing in pitch as it
pinged away into the distance. I stopped to let my heart settle back
to its normal speed. Through a clear window of ice, I stared into
an expanse of black water.

⋯ ⚎ ⋯

The Coeur d'Alenes are an Interior Salish tribe whose home range
encompasses the huge lake that bears their name. Its shoreline is
dappled with wide bays and generous wetlands that provide a rich
variety of muskrat habitat, and they have watched the rodents
adjust to snow, drought, ice, mud, famine, vigorous trapping, and
overpopulation. In 1927, an ethnologist recorded many old stories
told by Dorothy Nicodemus. Her daughter-in-law, Julia Antelope
Nicodemus, translated the tales into English, among them one

called "Muskrat Trespasses" involving muskrats and otters, family and territory.

> with muskrat little
> with his grandmother
> they had a house
> she said to him
> the little muskrat
> that old woman
> when in the morning then again she went to dig roots
> then little muskrat by his grandmother was said
> to this toward here
> eat about some grass
> not this toward here
> this toward here. . . .
> then he thought
> the little muskrat
> because why
> what is it that she forbids me?

When the story is read out loud, an English ear can absorb the Salish syntax to catch a familiar meaning: the more his grandmother told Little Muskrat to eat grass in one area, the more determined he was to eat somewhere else. He wandered onto turf that belonged to Otter, who punched him in the head repeatedly. When Little Muskrat got home, his head was so swollen that his little ears and eyes appeared even smaller than usual. Muskrat was angry, and against his grandmother's wishes paddled back to Otter's turf. He saw Otter's sister alone on the bank, strung his bow, and killed her with one arrow.

When the Otter brothers came to Grandmother Muskrat's house seeking revenge, she wrapped Little Muskrat's head in a poultice and put him to bed, then faced the wrathful Otters herself. She asked them why they would beat up such a small, helpless muskrat. All he could do was lie there, wondering who would do such a thing.

When the Otters left, Grandmother and Little Muskrat let out whoops of celebration. One of the departing Otters heard them and paddled back to see if he was being mocked. Grandmother and Little Muskrat quickly slipped down one of their holes and pulled a blanket over it, so that when Otter returned to their house he found

no one
in vain he searched it
no one
again he came out
hither he went around no one. . . .
there were just holes. . . .
he crossed to the other side of the water
just holes

When I took my next step forward on the ice, a blunt shape shot below me like a bullet. Through the ice I saw a single flash of limbs reach out and tuck smoothly back against a lean body as a lateral curl of tail steered the musquash on a graceful arc through the black water. The muskrat's stroke was instantaneous, but in no way hurried or desperate; the animal was simply swimming away. That is muskrat: root-storer; mound-builder; deft adapter to change; colonizer of new places; purveyor of dread diseases; cannibal and murderer; victim and survivor. A stream of bubbles drifted upward to kiss the bottom of the ice in regular, well-spaced intervals. That is muskrat, always slipping down another hole.

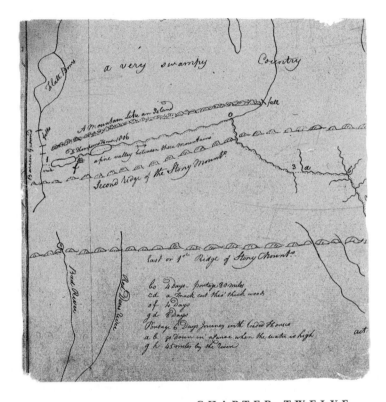

CHAPTER TWELVE

Reburying
Jaco Finlay

Map from Peter Fidler's journal, labeled *"Drawn by Jean Findley, 1806"*

Five Pipes ~

IN EARLY SEPTEMBER 1950, an archaeological dig along the Spokane River caught the imagination of the nearby city. On a low rise just above the floodplain, investigators hoped to find the site of one of the first fur trading posts in the region. The workers shoveled into a promising trench so quickly that a local reporter wondered if the crew didn't have "a little black magic" going for them; in fact, tribal people with knowledge about the post had pointed to exactly the right spot to begin digging. By the following summer, patient excavation had outlined the overlapping footprints of Spokane House during its occupation by three successive fur companies between 1811 and 1825.

While trying to locate a bastion at the corner of one stockade, workers exposed several large flagstones a few inches beneath the surface. Below the stones they found a thin slab of rotten wood covering a shallow pit and remnants of a wooden coffin edging the hole. The heavy flagstones had collapsed the coffin's lid, damaging the human skeleton below; the processes of natural decay had almost completely decomposed the ribs, leaving little more than an empty rectangle framed by relaxed arm bones, a dented skull, and a flattened pelvis, with crossed legs trailing off below. Analysis revealed that the remains belonged to a mature male with an enlarged right tibia, probably the result of chronic rheumatism.

Thirteen nails, a few tatters of cloth, and three brass buttons were found inside the rotted coffin. A number of personal belongings had been buried alongside the body as well: a fragment from a bone comb, a tin drinking mug, and a disintegrating hunting knife sheathed in a thin metal scabbard. A three-by-ten-inch portion of writing slate, together with the nose piece and broken lens from a pair of spectacles, indicated that the person in the grave was educated, and an assortment of pipe bowls suggested a certain fondness for smoking. There were five in all: two pipes of clay, one carved from wood, a stone bowl that still held the lead ring used to attach a stem, and a copper one beaten smooth.

One of the clay pipes had four straight lines scratched into its bowl. Two of the lines formed a fairly distinct letter J, and a pair of horizontal nicks flew off the vertical. Local historians were convinced that the cryptic "JF" was the monogram of none other than Jaco Finlay, an early furman who had been living at Spokane House at the time of his death. Local legend regarding his burial at the old post was bolstered by an entry in the journal of Nathaniel Wyeth, who passed by the abandoned trade house in 1833 and noted that while most of the buildings had been burned for firewood, one bastion had been left standing out of respect for a dead clerk who was buried beneath it. Within days of the find, a newspaper article trumpeted, "Discovery at Spokane House: Jaco Finlay and His Grave."

In the summer of 1952, the ongoing survey uncovered another burial site containing the remains of a woman, a girl, and a man, all bent forward at the waist in the traditional posture of Interior Salish interments. Alongside the male skeleton, the diggers found three trade muskets and two short decorated wooden bows. The stocks of the trade guns had been broken off, and a copper kettle found nearby had been bashed in; the chief archaeologist noted that this destruction was in keeping with a Salish practice of "killing" articles to be buried with a corpse. Other items unearthed at the grave site included a quantity of small blue beads and some dentalia shells. The artifacts were shown at a public museum in downtown Spokane, and museum interpreters began to ponder a more permanent exhibit. In a time when native remains were commonly displayed, someone broached the idea of leaving the crouched skeletons in situ and peeling away enough dirt so that the graves could be glassed in for view by a walk-through audience. "With proper lighting and glass boxes, this could be done in a respectful and educational manner," one report concluded.

Jaco Finlay's remains had been removed from the excavation site the previous year and placed in storage, but no one suggested including them in the display. Speaking in October 1953, the president of the local historical society admitted, "We are somewhat at a loss to know what to do with Jaco, but at this time his resting place is a box in our Eastern Washington State Historical Society Museum, awaiting some program of improving the Spokane House area, when Jaco's remains should be marked with a proper marker near his original resting place." If the museum wasn't sure exactly how to handle Jaco, at least they knew where he belonged.

Jacques Raphael ~

JACO FINLAY'S FATHER, James, was born in Scotland in 1734 and shipped to lower Canada when he was around thirty years old. There he fell in with the array of independent Scotch, French, and New England traders who eventually coaslesced into the North West Company. By 1767 James Finlay had wintered in the "Indian Country" beyond the Great Lakes and produced a son with a Montreal wife. Before the next winter blew in, he had established the Fort Finlay trade house downstream from the Forks of the Saskatchewan River, well over a thousand miles west of his urban family. At his field post, Finlay took a mistress from the Saulteaux tribe, an Eastern Woodlands people who had absorbed elements of both fur trade and Cree culture as settlement pushed them north and west from the Great Lakes. Around 1768, James's unnamed "country wife" bore a mixed-blood child they named Jacques Raphael. In time he acquired a nickname that was pronounced in the French way, Jocko, but spelled by the most consistent writers of the time as Jaco.

Over the next decade, James Finlay continued his seesaw life, journeying back and forth between the city and the wilderness and fathering six more children by his wife in Montreal. In 1776 he was still manning his namesake trade house, a sprawling and prosperous post on the Saskatchewan. Jaco would have been eight years old then, the country son of an important feudal boss. He would have watched brigades of voyageurs moving up the river as the fur companies pressed farther and farther westward, and he would have been about twelve or thirteen when the smallpox epidemic of 1780–81 wreaked its terrible havoc. By then James Finlay had already begun to reduce his investments in the fur business, and by the mid-1780s he had received an appointment as the official Inspector of Chimneys in the city of Montreal, where he died in a comfortable two-story house in 1797.

While there is no record of Jaco ever joining his father in town, James Sr. apparently made certain provisions for his mixed-blood son. The youngster must have learned the basics of reading and writing because, along with his half-brothers, he was commissioned into the North West Company as a clerk. But instead of regularly returning to lower Canada like his father and siblings, Jaco's path opened West—upstream on the Saskatchewan, across the Prairies, to the Front Range of the Rocky Mountains.

His first appearance in the fur trade record occurred when he was twenty-six years old and stationed at the North West Company's Upper Bow House (near modern Medicine Hat, Alberta). In June 1794, the Hudson's Bay Company establishment a short distance away came under attack by a raiding party, who killed men, women, and children, pillaged all the goods, and put the buildings to the torch. The attackers then moved on to the North West Company post, where Jaco Finlay organized a handful of men to fight off the assault. When one of the raiders stood up to urge his cohorts on, Jaco shot him dead, and the attackers retreated. One account credits Jaco with rescuing the sole survivor of the neighboring post, who was hiding in a cellar beneath the wreckage of the burned fort. North West partner John McDonald of Garth, recounting the event years later, said: "Our fort was in charge of one Jaccot Finlay, a man of courage."

Two years later, Jaco was posted on the North Fork of the Saskatchewan, where he apparently manned several posts for the remainder of the decade. In 1799, while his Montreal-born half-brother John was promoted to head the whole Athabasca District, Jacques Raphael Finlay appeared on the pay list as a clerk receiving twelve hundred livres per year (one French livre was worth about seven-eighths of an English shilling), one of the highest wages on the chart and exactly equal to that of veteran fur agent and surveyor David Thompson, recently hired by the North West Company.

While nothing is known of Jaco Finlay's activities during the subsequent five years, the company's 1804 pay list retained him at the same level he had held in 1799—a person of mixed blood could not aspire to any position higher than clerk in that era. It is likely that he continued to ply the upper Saskatchewan as the company added new posts farther upriver in pursuit of trade. He had a native wife by then, often identified with his mother's Saulteaux-Cree tribe, and at least three children. In the fall of 1806, he became the point man for the North West Company's corporate expansion across the Rocky Mountains. Jaco Finlay's knowledge of the country and his facility for native languages made him a logical choice as chief scout for the adventure. John McDonald of Garth, the agent at Rocky Mountain House and one of the prime instigators of the ambitious plan, sent Finlay ahead to clear trail and lay up two canoes for the party that would come across in the spring.

By October, Jaco had crossed the Divide along with his family and some of his in-laws and was already at work building canoes. Late in the month he sent two of his helpers back across the mountains with a young Kootenai to fetch more supplies and tobacco. In mid-November Jaco himself visited Rocky Mountain House for a week, and it may have been then that he drew a rough map containing much accurate information about drainages and trails on both sides of the Rockies. This sketch later appeared among the papers of Hudson's Bay Company agent Peter Fidler with the inscription "Drawn by Jean Findley, 1806"—a name close enough to that of the only known person who had the information needed to draw such a chart.

In late February 1807, Jaco again sent a courier from west of the Divide to fetch sugar, flour, and spirits from Rocky Mountain House; at the end of March he and his wife and two children all returned to the fort. When word reached the trading post on April 20 that a group of Kootenais had crossed the pass with furs to trade, Jaco and a young clerk named Finan McDonald went out to greet them with tobacco.

Jaco's Campment ~

WHEN DAVID THOMPSON LED THE COMPANY'S expedition across the Rockies in June 1807, Jaco Finlay was not with them. The party ascended what is now called Howse Pass and then staggered down the west slope of the Rockies along the Blaeberry River, which was running wild with meltwater. As they proceeded, Thompson became increasingly unhappy with the work of his advance scout. The path had not been cleared wide enough for packhorses, and although Jaco had assembled the parts for two small canoes, "hardy enough for light voyaging," he had failed to cache the proper timbers for a cargo canoe, gum for caulking the seams, or suitable birch bark for sheathing the boats. "By this vast trouble we have had, it may be seen how Jaco [and his men] . . . have earned their wages. Those of Jaco ought certainly to be forfeited as he has done next to nothing," Thompson wrote in a report to the North West Company partners.

While he overwintered at Kootenae House, the trading post he built at the source lakes of the Columbia, Thompson's anger began to cool. In a letter written in March 1808 he referred to the trail-clearing fiasco, but went on to admit that his own incomplete knowledge of the country had contributed to his problems, and he left open the possibility of working with Jaco again—he was the man, after all, with the most experience in the region. "Jaco has behaved like a scoundrel," Thompson wrote, "but all is now over." Such forgiveness did not prevent him from taking a jab at Finlay in another letter written the next day. Apparently Jaco had gone down to the Bow River country to trap and trade, and Thompson "could not help but smiling" at his projected returns. "I will bet any of you two to one of a whatever you please that he does not get 10 packs from them in 2 years. As to our being friends or enemies that will depend on the interest of the concern."

By this time Jaco had given up steady employment as a clerk for the life of a free hunter, occasionally working for the North West Company while trapping independently the rest of the time. He often traveled with tribal and mixed-blood trappers, any of whom might take on temporary employment with the company as scouts, hunters, or interpreters. When David Thompson's Columbia brigade passed down the Saskatchewan in June 1808 on their way to deliver the year's furs to the company warehouse on Lake Superior, they found Jaco camped near Rocky Mountain House along with some of his in-laws. Thompson paused long enough to pick up Finlay's winter take of sixty-seven beaver pelts, one otter, and a pair of muskrats—about one fur pack. Thompson's daybook does not mention whether this meager haul provided him with another smile, but the two men continued to work together, off and on, for the next five years. Since Thompson left behind a wealth of field journals and Finlay left none, literally eveything we know about Jaco during this period comes through the pen of David Thompson.

It was five months later that Jaco next appeared in Thompson's daybook, arriving at Kootenae House on a snowy November afternoon with his family and five Iroquois men named Joseph, Pierre,

Ignace, Martin, and Jacques—a cadre of experienced trappers whom Thompson had recruited to trap on the west side. After Jaco and the Iroquois had settled into camp a short distance from the post, clerk James McMillan paid them a visit to settle accounts for the eight and a half packs of furs they had procured over the summer. Some of the payment was in the form of rum, and after business was concluded, the group "drank and fought all night." Next evening, McMillan returned to Jaco's camp for another round of revelry. During the rest of the winter, Finlay remained close by in the Columbia Valley along with his Iroquois compatriots and his wife's Saulteaux-Cree relatives.

At some point Jaco took an interest in the herds of wild horses that grazed in the fine grassy hills above the Columbia's source lakes. According to Kootenai elders, these were the feral offspring of horses that had been turned loose after an epidemic of smallpox killed their owners some years before. The voyageurs called the horses *marrons*, a French slang term for something tame gone back to the wild. In early January 1809, Jaco visited Kootanae House to purchase a new gun and reported that he had captured and halter-broken a string of ten marrons; a few days later his count had reached eighteen. Within the week, Thompson joined in the chase, and he later described a delirious day of riding full tilt over the hills, running down the spirited mustangs.

That spring, the wife of a Kootanae House voyageur suddenly passed away, leaving four small children, the youngest only six months old. A week later Thompson paid a visit to "Jaco's Campment" and left the children in the custody of Jaco's family. It must have been Finlay's anonymous wife, who already had several small children of her own to care for, who accepted the responsibility. While he was there, Thompson also ordered two new canoes and asked Jaco to secure provisions for his men in preparation for their journey east to deliver furs. The willingness of the Finlays to care for four orphaned children did not keep Thompson from criticizing Jaco's boatbuilding skills, however. Only a few days later, the explorer complained that the canoes were far behind schedule and so badly made in the bottom that they would never last.

That summer Jaco's camp received a much rougher visit from marauding Piegan Blackfeet, who made off with all his fine marrons and most of his other property as well. The ravaged family was ascending Howse Pass on foot when they met the North West Company brigade returning from the east. Upon learning of their plight, Thompson supplied the Finlays with horses and other necessities, and the family reversed directions to accompany the voyageurs on a journey down the Kootenai River through what is now western Montana and northern Idaho. Jaco speared fish for the furmen along the way and brought them meat as they were thatching the roof of their new Kullyspel House on Lake Pend Oreille. The furmen were among Interior Salish–speaking people now, a loose collection of tribes who maintained a relationship with the Kootenais and were sworn enemies of the Blackfeet.

Jaco and his family moved east to Salish winter camps in the dependable grasslands along the Clark Fork and Flathead Rivers. After exploring to the west, Thompson followed his scout's footsteps and decided to construct another trading post called Saleesh House in close proximity to those camps. But with winter closing in quickly, and no one having any luck at hunting, the voyageurs became so weak with hunger that they could not pursue their work on the post. Thompson's journal entries took on an air of hungry concern as more gameless days drifted by, until Jaco and his wife arrived bearing dried beaver tails and pounded meat that replenished the men's strength.

Jaco may well have spent most of the winter with a Salish band not too far upstream, in a beautiful small drainage now called the Jocko River. The following spring, 1810, Thompson put Finlay back on the company payroll "in his old Capacity as Clerk and Interpreter." Soon Jaco and his family, trailed by several familiar

Iroquois, arrived at Saleesh House to help wrangle the spring brigade of fur packs downstream. Before departing on his annual trek east, Thompson gave Jaco instructions to travel to the Spokane drainage and supervise the building of a post to serve the people there.

Finlay chose a wedge of land formed by the confluence of the Spokane and Little Spokane Rivers, the site of an important village of the Middle Spokane people. The salmon and steelhead fishery at the spot attracted a wide variety of visitors and made a natural location for a new trade house. Jaco oversaw the construction of the post and, with some help from Finan McDonald, managed its operations until June 1811, when David Thompson arrived to enter Spokane House into written history: "Thank Heaven for our good safe journey, here we found Jaco etc. with about 40 Spokane families." Over the next ten months, Jaco's trading post served as a hub for exploration and transport—the center for horses purchased and pastured, messages sent, new men left off, old hands redeployed. When Thompson returned that fall from his epic trip to the Pacific Ocean, he dropped off a Hawaiian voyageur named Coxe at Spokane House so that Jaco could mentor him in the trade.

Jaco apparently remained at Spokane House during the winter of 1811–12, and it may have been around this time that he married a Spokane woman called Teshwentichina. In the spring of 1812, he was summoned to the Colville Valley, where David Thompson was building cargo canoes. There the two men, whose relationship spills over any attempt to qualify it—peers and rivals; comrades and explorers; boss and laborer; pen man and point man—began their last collaboration. Thompson oversaw the construction of two cedar bateaus while Jaco took charge of a pair of birch bark canoes. Boards were knifed and split, bent and sewed. Thompson's journal recorded dogged progress on the boats for the next nine days, and at each phase, Jaco lagged a little bit behind. As Thompson began to split out oars, "Jaco's second canoe the side seam sewed only." The cedar plank boats were both being dragged toward the

Columbia before Jaco had the first of his birch bark canoes completely timbered. But by April 22, all four vessels were in the water,
crammed with over nine thousand pounds of furs, ready to depart
on their three-hundred-mile trip upstream to the mountain
portage. David Thompson, whose path had crossed and recrossed
Jaco Finlay's for thirteen years, stepped into a canoe bound for
Montreal and paddled out of his life forever.

<hr />

Within a month of Thompson's departure, Finlay had introduced
himself to the Pacific Fur Company, an American venture that was
expanding its business upstream from Astoria. An agent at Fort
Okanogan was told that if a Mr. Jacques Finlay sought trade goods
on credit, he should be accommodated with everything except
liquor and treated well. A sales slip from the company records indi-
cates that Jaco did indeed visit the American post:

> Received of Mr. Donald McGillis, the following Goods—8 Half
> and 4 Small, etc. etc. to lay out in the Indian trade for the interest
> & good of the P. F. C. [Pacific Fur Company] duly received
> by me
>   (signed) Jacque Finlay

The signature on this chit preserves the only known stroke of
Jaco Finlay's hand. He carried it out with apparent relish, adding a
fat double flourish to the final Y, which swings back across several
previous letters. It is the autograph of a man who was comfortable
with a pen.

As for switching allegiance in an international fur trade rivalry,
Jaco surely considered himself a free hunter at heart, entitled to
play in any new game that came along. This particular clash
between fur interests was of little consequence anyway, vanishing
when a British man-of-war rode into the mouth of the Columbia

on the eve of the War of 1812. The North West Company purchased all Pacific Fur Company holdings outright, then immediately resumed its role as lord of the Columbia District. The partners apparently held no ill will toward their wandering clerk or his sons, because the company's employment list for the winter of 1813–14 contained no fewer than four Finlays:

53. Finlay, Jac. Rap   Clerk & Interpreter
54. Finlay, Rap., Jun.   Interp. & Hunter
55. Finlay, Thorburn   M[ileu] & Hunter
56. Finlay, Bonhomme   M[ileu] & Hunter

Jaco's contract at Spokane House extended until April 1816, and his wage was listed as eighteen hundred livres, six hundred more than other established clerks like Finan McDonald and six times that of his son Thorburn, who was hired as a *mileu* (paddler in the middle of the boat) and hunter.

After the Hudson's Bay Company juggernaut absorbed the North West Company a few years later, Jaco's extended family continued to range all over the Columbia District in various capacities. Two Finlay sons accompanied a Snake River expedition to the far reaches of the Boise Basin, and another manned a boat at the mouth of the Columbia. Others intercepted Shuswap trade at the Canoe River, carried furs across the Athabasca Pass to Jasper House, eased south into the Willamette Valley and north to the Fraser River. Sons and daughters married into families in the Flathead Valley, in the Pend Oreille marshlands, and on the Spokane prairies. They nestled into the close confines of the Colville Valley and joined in the annual Salish rounds of fishing and root digging, fur hunting and horse racing.

While his family spread far and wide, Jaco apparently stayed close to his old haunts in the Spokane country. In 1825, the Hudson's Bay Company moved their operation from Spokane

House north to the new Fort Colvile near Kettle Falls, and after the last iron hinges had been stripped away, Finlay took over the husks of the buildings. No record has survived of any committee decision that this was a fair reward for services rendered; there is no trace of money or goods changing hands. But it seems fitting that the man who had originally chosen the site now stayed behind as its caretaker.

A Morsel of Food ~

Jaco Finlay was at home at the old Spokane House on the spring morning in 1826 when David Douglas came to call. The botanist had ridden down from Fort Colvile with two of Finlay's sons, covering eighty wet, chilly miles at a rapid clip, and they arrived at Jaco's place just before noon on their second day out. "Mr. Finlay received me most kindly," Douglas recalled, "regretting at the same time that he had not a morsel of food to offer me."

Jaco told Douglas that his family had been subsisting for the past six weeks on camas bulbs, bitterroot, and "moss bread" derived from "a black Lichen which grows on the Pines." Since Jaco's wife Teshwentichina was a Spokane, there was nothing unusual about having such traditional fare around the house. Douglas was not familiar with the local method for transforming the lichen into edible cakes, and thus a good portion of his account was given over to a recipe for steeping and roasting the lichen bread.

> The mode of preparing the latter is as follows—after clearing it thoroughly from the dead twigs and pieces of bark to which it adheres, it is immersed in water, and steeped till it becomes perfectly soft; when it is placed between two layers of ignited stones, with the precaution of protecting it with grass and dead leaves, lest it should burn. The process of cooking takes a night,

and before the lichen cools, it is made into a cake, much in the same way as the Camas; when it is considered fit for use. A cake of this kind, with a bason of water, was all that Mr. Finlay had to offer me.

Douglas did not actually observe the preparation of the cakes, and according to tribal members, his recipe omitted camas and wild onions, which the Spokanes always mixed with their lichens. Perhaps not as eager to eat these cakes as he was to describe them, Douglas dug some dried buffalo meat from his saddlebag and shared some fresh game—possibly one of the long-billed curlews he had shot along his journey. When it was all spread out, the two men sat down to the business at hand: Douglas's musket was knocked up, and the agent at Fort Colvile had recommended Jaco Finlay as the only person in the entire country who could put it to rights, and had added that he was also just the man to provide extensive information concerning local natural history.

After lunch, Douglas "hastened to inform him of my request, though my imperfect knowledge of French, the only language that he could speak, much limited our intercourse, and prevented my deriving from him all the information that I wished to obtain." The botanist left his musket for repair and spent the afternoon on a long collecting walk up the Spokane River. Upon his return he was delighted to find that "Mr. Finlay had obligingly put my gun in good order, for which I presented him with a pound of tobacco, being the only article I had to give."

Douglas must have made some progress with his French, because he spent two more days botanizing around Spokane House, delivering everything he found back to the patron of the house for inspection. Particularly taken with the local varieties of currants, he was delighted when Jaco explained how three different blossoms produced three distinct berries. The naturalist engaged Finlay to collect specimens and seeds for him, and asked him to be on the lookout for a particular kind of onion with a delicious nut-like root.

As their halting conversation turned to animals, Finlay seemed to recognize a peculiar local species of mountain sheep that Douglas had heard about, and promised to bring back a *Mouton Gris* when the family was out hunting in the fall. While Douglas was snipping at the local plant life, one of Jaco's sons brought down a grizzly bear in the hills above the river and presented it to the collector, but the animal was too large to be preserved. Clearly delighted with the idea of a crew gathering new specimens, Douglas gave the Finlay sons a lesson in the way he liked to have his skins dressed and promised to pay them if they would procure different animals for him.

One of the sons escorted Douglas back to Fort Colvile, and later that summer Jaco's oldest son James guided the botanist on an excursion into the Blue Mountains. At Fort Walla Walla, Douglas hooked up with John Work and accompanied his pack train back to Fort Colvile. When the party crossed the river near the old Spokane House, Jaco Finlay emerged to offer his guests some salmon freshly taken from his weir in a branch of the river. Douglas's journal offers nothing more than the simple meal of fish, and that is the last morsel of Jaco Finlay we have to chew on. He died at Spokane House in May 1828, approximately sixty years of age. John Work heard about Jaco's passing while he was hauled out downstream on the Columbia, trying to repair a damaged cargo canoe.

Over the summer, news of Jaco's demise would have traveled the river, touching his many children and relatives, plus the clerks, voyageurs, free hunters, and tribal members who had smoked with him during his time. In early October, two Saulteaux-Cree brothers from west of the Divide arrived at Edmonton House, bringing news of Finlay's death. The siblings could have been the same in-laws who had helped Jaco blaze the Howse Pass trail for David Thompson in 1806. The pair traded for a few supplies and took off again the next day; having reported from across the divide of time, they disappeared once again into the Columbia country, just as quietly as Jaco had so often during his long run in the trade.

Grandmothers ~

IN THE FIRST SUNSHINE of early May, I stood at the railing of a small bridge over the Colville River. The water roiled through a sprawl of red-osier dogwood, hawthorn, alder, and clumps of wild rose. On the face of a freshly eroded bank, the winding tunnel of an old muskrat den was clearly visible. Black-headed grosbeaks were already singing hard, and within the next few weeks, as the bushes leafed out, orioles, catbirds, chats, and redstarts would arrive to fashion their nests among the tangle. The Colville Valley forms one focal point of the Finlay family story, and it also provides six generations of time for watching the river flood and dry up; for its banks to get stripped clean and grow back; for Finlays to work as miners and farmers, postmistresses and schoolteachers, to move away for a while and then touch base again. This was how I imagined the Colville Valley must have looked and sounded during the many spring traverses made by Jaco Finlay and his children.

Local historian and raconteur Walt Goodman stood beside me on the bridge and considered the notion.

"My mother, now," he said, "she could remember the time before the river was dredged, when the water would sweep around here, with a current slow enough that you could dawdle around in a rowboat, and pools deep enough to hold big trout. She used to walk down behind her house in the late afternoon—her dad's farm was right over there—and be back with a few nice fish for supper. I expect my grandmother did the same thing, and Mary King, my great-grandmother, before that."

Mary Ann King was the wife of retired Hudson's Bay employee Peter King. She was Walt's great-grandmother and Jaco Finlay's granddaughter, a connection not uncommonly traced and honored among local inhabitants. Mary King would have eaten plenty of curlew during her time. She would have seen muskrat push-ups spread across the entire valley, and she might have been able to say whether it was her father, Patrick Finley, who once guided David Douglas through the marshland.

Walt pointed to a farm that hugged an eastern twist of the valley floor less than a mile away.

"That was probably Patrick Finley's place," he said. "He was my connection, Mary King's dad. Other brothers were scattered around nearby, shuffling from this place to that. I think most of the boys liked to move around. But not all of them." Walt reminded me that when Elkanah Walker and Cushing Eells opened a mission at Tshimiakain in 1839, they soon became acquainted with Patrick Finley, whom Walker called "Pishnot," his three brothers, and a sister living in the Chewelah valley. On journeys to and from Fort Colvile the missionaries would regularly spend the night in the cabin of Jaco's daughter Josette and her husband, a Hudson's Bay Company employee.

I asked how many children Jaco had, and Walt, who took rightful pride in his clear mind, took a stab at the lineage, but was slurring and grinning long before he could spit out the whole list of Jaco's offspring, somewhere around eighteen strong. However neatly he tried to rattle them off, they never fell out exactly the same. "And all of Jaco's children spelled their last name *e-y* instead of *a-y*," Walt said. "Why do you think that was? I asked my granny about that one time, when I was first getting interested in names, and all she did was turn away."

Walt turned his head to mimic the way his grandmother avoided the subject. "She just never talked about that part of her background—a lot of people back then didn't want you to know they were mixed-blood, Metis, whatever you want to call it. I don't know if I ever heard her say a word about Jaco or his children."

Still, it is well known that several of Jaco's daughters married Hudson's Bay Company men, and many of his boys appeared on the pay lists of contemporary fur brigades. The many Finlay and Finley entries in the fur trade journals amount to brief glimpses of the masculine line—hardly enough to keep track of a whole family, but when Jesuit missionaries arrived in the upper Columbia at the end of the 1830s, Jaco's legacy began to take shape on paper. Throughout the network of small Catholic missions across the Northwest, his children's and grandchildren's names were recorded in parish ledgers of birth, baptism, marriage, and burial. Although Jaco was identified on one daughter's marriage record as "Finlay, infidel," it was the infidel's progeny who were attending the mission churches.

Jesuit Father Pierre DeSmet rode down the Colville Valley in the summer of 1845, and after receiving permission from the local Salish people, he established St. Francis Regis Mission to the Crees beside a cottonwood spring. Several Finlays attended the services there, and when DeSmet returned two summers later he reckoned that "already about seventy Metis or half-breeds have collected to settle permanently." It might have been during that second visit that Father DeSmet, always curious about genealogy, drafted a family tree on a piece of foolscap paper. The name "Jaco Finly" climbs a stout genealogical trunk. Fifteen branches spread elegantly from his stem, each one tagged with a son or daughter's name: James, Josette, Augustin, Pichinna, Kiakik, Jennessie, Francois, Jaco Migwham, Isabelle, Nicolas, Baptist, Marguerite, Joatte, Rosette, Basil.

Pichinna, the second limb from the bottom, was Walt Goodman's third grandfather, whom he knew as Patrick. "So many languages," said Walt. "So many sounds. Patrick; Pichinna; Bish-e-nah. They're all the same guy." In Father DeSmet's rendering, Pichinna's loaded branch has to curl back on itself to avoid running off the page; his family numbered four wives and sixteen children, who had already blessed him with thirteen grandchildren.

The second and third generation of Finleys, the branches and twigs on Father DeSmet's tree, continued to spread across the Northwest. When the international boundary was established in 1846, they kept moving back and forth across it. When a measles epidemic led indirectly to the Whitman massacre in Walla Walla, Patrick's brother Nicolas was in the thick of it. When gold was discovered in California, various sons and daughters packed down to the Sierra Nevada, and returned with placer-mining knowledge that resulted in an early strike in Montana. When a prospective settler named John Campbell rode north through the Colville Valley in the late winter of 1855, he met four of Jaco's sons and their offspring firmly entrenched in a community of former Hudson's Bay Company workers. But their sister Josette was no longer with them; she and her husband had moved to Oregon and taken up a homestead on the Umpqua River.

＋・ ≡◆≡ ・＋

Walt Goodman watched a root clump come twisting down the Colville River and disappear beneath the bridge. He crossed to the other side in time to watch the mass bounce out and roil away into the hawthorns.

"Always riding," he said. "Lots of horses running through these stories. I heard one, I think I can remember it, about a Finley who was traveling from Fort Colvile down the river somewhere, to look at a horse—this brother was a great one for trading horses. He was going to Wilbur I believe, that's quite a ways, and for some reason he had to carry a child there as part of the deal. Now this Finley brother always rode bareback. He put the boy in front of him, astraddle the horse's withers, and took off.

"They say this Finley was tall and jawboned, kind of a hard man. The boy who was with him didn't fit across the horse's neck quite right, and that made the horse shy, so that it kept flicking its skin—you know how a horse twitches to get the flies off? Every time it flicked, the boy would cry. The horse kept flicking, and that boy kept crying, but no way was Finley going to stop. They had themselves a long, miserable ride down to Wilbur.

"I don't think any of the boys stayed in this valley very long at one stretch. They all died over in Montana, you know—Patrick was buried in Frenchtown, a horse accident if I'm not mistaken, in 1885. But some of the women stayed behind. You know what happened to Mary King."

In 1910, Mary King and her children applied for official enrollment in the Confederated Colville Tribes, which would entitle them to an allotment of land and a small amount of money. Initially, the council voted to include Mrs. King and some, but not all, of her children. The following year, the council reversed its decision on the grounds that Mary King was one-eighth Cree. In November 1912, Walt's great-grandmother and several of her daughters appeared before the council at the Colville Agency in Nespelem to appeal their applications. In order to be approved, Mary King had to demonstrate kinship with the Colville Tribe.

When Mary Ann King stood before the council, her plea was read by Superintendent J. M. Johnson: "This old woman says that she has lived at Chewelah all her life." Mary King didn't know exactly where she was born, but her parents had been employed by the Hudson's Bay Company, and her father was Patrick Finley, who she thought was half Spokane. Her mother spoke French, lived most of her life in the Colville Valley, and died in Montana.

Fur trade descendant George Herron came forward to witness. "I knew Mary Ann King," he said, "and knew her father. His name was Bish-e-nah. He was one of the Finleys from Red River. . . . He talked several languages, Spokane, Pend Oreille, Cree, and French. He was a half-breed. When I first saw him in Colville he was married. He continued living in the Colville Valley at Chewelah for a while and then went to Flat Head and never came back. Mary Ann, this woman, never left."

Joseph Grand Louie said that he knew Bish-e-nah and his wife, and that both of them had spoken French, Cree, and Kalispel but not the Colville language. He did not know their tribe, but supposed it to be Cree or Kalispel.

Mary King protested that she could speak a little Colville, and understood Colville very well. She said she knew the Colville chiefs. Superintendent Johnson spoke again, affirming that Mrs. King's four children had attended the Colville Mission School. Then he said, "Unless there is something more to be said I will ask for a vote on the case now." In regard to her land allotment, Mary King received twenty-three favorable votes and sixty-three unfavorable; as to money, fifteen favorable and sixty-eight unfavorable. In an official document, the reason for her refusal was listed as "too much Cree."

Sometime around 1900, an itinerant photographer took a picture of Peter and Mary King's homestead on the open floor of the valley. In the center of the frame, a handsome house is fronted by a sharp picket fence and a line of solid-looking men, all well turned out in a mix of northern European and native dress, with a touch here and there of the cowboy. The way the light falls on the house behind them accentuates the perfect dovetail joints at its corners.

Up on the porch stands a lone woman, almost hidden in the shadows. Her face is dark and full of character; she looks older than any of the men and is the only obvious native in the scene. She wears a long black dress that fastens close around her neck and sweeps the rough wooden planks of the porch floor. Her hair is parted down the middle and pulled back tight to merge with the blackness of her dress. She is doing something with her hands in front of her, perhaps holding an object, but the darkness of the photo makes it impossible to tell.

"Mary Ann King," Walt said, pointing to the figure in the

shadows. "Patrick Finley's daughter. She's standing on her own porch in the house that her husband built. They say he was a really fine blacksmith and carpenter. He would have been gone, what, fifteen years at the time this picture was taken. Mary, now, she was a good-sized woman. Quite a hand at preparing hides, scraping and curing deer, beaver, things like that.

"I only saw her three times in my life that I know of. Once at the house that's in the photograph, when I was too young to remember much of what went on. My older sister used to spend the night there to take care of her. She'd cook for her, rub her back to ease her sore muscles. Then when Mary's health started to fail, she moved to Priest River with her daughter. We visited her there before she went to a nursing home in Idaho. One time I remember was in 1925, I was ten, and she wasn't doing well at all. I don't know if she was sick or what, but it was kind of scary for a kid. She was lying on her back, and all she would say . . ."

Walt stopped himself for a moment, studying the memory.

"She didn't speak English so well, you know. She looked at us and said, 'Me poor old woman.' That's all she'd say. 'Me poor old woman.' "

"It's funny," Walt said after a while. "I don't really have anything to remember Mary King by, but when I see her in the photograph I think of this day I spent at an Indian fair. It always puts her in my mind. Seems like a crazy thing, because it didn't happen until 1938, and she was long dead by then.

"The civic club I was involved with raised enough money to buy some carnival rides—an old merry-go-round and one of those hubs that little airplanes spin around. We had a popcorn machine and a cotton candy machine, and we started taking them around to little towns on Saturday. The people on the Colville Reservation at Inchelium got wind of it and really wanted us to bring the show to them.

"We got the rides set up and everything seemed to be going

OK. Then I noticed this older Indian woman standing nearby. She was talking, not to anyone in particular, and I figured she must have some grandkids on the merry-go-round. I thought that was nice. But when I stepped a little closer I could hear her whispering 'Goddamn' in this angry way. 'Goddamn you, Blue Eyes,' she said. She repeated it over and over, really soft. 'Goddam you, Blue Eyes.'

"I don't know what could have gotten that lady so ticked off, but I think about her all the time. They say Jaco Finlay had blue eyes. I heard the same thing about Patrick, that they used to call him Blue Eyes." Walt touched the top of his head, as if he was surprising himself. "And I don't know why, but that lady at the carnival always makes me think of my grandmother Mary Ann King."

Honorable Ancestor ~

"I NEVER COULD GET ALL THOSE SONS STRAIGHT," said Jeannette Whitford from her living room chair. "Jaco, James, John, Jacques, George—they all sound the same to us. But my grandfather's father was one of Jaco Finlay's sons. Old Finlay, I believe he lived a lot like an Indian. We call him Honorable Ancestor."

Jeannette was born on the Coeur d'Alene Indian Reservation, even though her parents were enrolled Spokane. "What the white people call Middle Spokane, those are my people. They fished at the mouth of the Little Spokane, and told Jaco Finlay to build the first Spokane House near their village there. They ran horses on Peone Prairie to the north. The women dug roots. My family ended up on the Coeur d'Alene Reservation because the Jesuits had baptized us as Catholics."

The Spokanes who were removed to the Coeur d'Alene Reservation tended to stick together. Jeannette's grandfather died when she was four, and her only sister was thirteen years older than she was. Maybe that's why she got into the habit of sitting with the Spokane elders whenever they gathered in someone's house to talk. When her parents had a disagreement about whether she should go to the mission school at De Smet, she was happy to remain at home, listening, for a couple more years. The old ways of talking and storytelling became ingrained in her memory, and she grew fluent in two languages.

"That girl, she can listen," the older Spokanes would say. "She knows what names to call us—who is grandmother, and aunt, and uncle. She knows Indian."

Seated at the feet of her elders, Jeannette heard the stories that took a long time to tell and were full of the nuance of languages spoken by the separate bands. She heard all about the point of land formed by the joining of the Spokane and Little Spokane Rivers, and the sustenance it provided for the Middle Spokanes.

"There was a village there for many generations," she said. "The people stored bitterroot and camas in pits on the Little Spokane side. There were caves somewhere around where a golf course is now that held ice all through the year, and that's where they would keep their fish. The elders talked about those places.

"Sometimes they would talk about two brothers, sons of Jaco Finlay. These two liked to travel together; they were real close that way. I was never sure exactly which ones they were because when the elders would sit around and say 'cousin' and 'aunt' and 'uncle' it depended on who had died, and who was left, and who you were already related to. The elders knew all of that, but I think a lot of people who were writing the names down didn't understand and got confused.

"But there were these two brothers who were always together, and they called one of them Blue Eyes because of his eyes. The other always rode without a saddle, so they called him Rode with Just a Blanket. In our language that sounds like the word for breechcloth, but it isn't. These two brothers rode all over the country together and visited everyone, from the Flatheads to the Colvilles to the Coeur d'Alenes. They must have known their manners and what to call their aunts and uncles, because everywhere they went they were welcome. I can't tell you their names or what they did or who their children were. I don't know exactly who their wives were. I only know that they were Jaco Finlay's sons, and they were welcome.

"I think about those women of the tribes, though, and the furmen they married. What did the women think about them? You know white men, that sour smell they always carry with them. How could they stand it? When one of our men had bad body odor like that, we would take him out and bathe him in the earth of a freshly dug molehill."

Jeannette laughed at the tradition, and at herself for getting so worked up. She had grown up that way, aware of both the worlds around her. She attended the mission school at De Smet for four years, then switched to the public school in Worley, then went through university. In time she won, then lost, a seat on the Coeur d'Alene tribal council. When representatives asked her to serve as judge for the tribal court, she accepted the responsibility and set up a juvenile justice system.

In the early 1970s, the Eastern Washington Historical Society offered her a position on their museum board. There Jeannette Whitford listened to stories about the excavations at Spokane House. She looked at the artifacts that had been unearthed, and after a while she realized that bones identified as those of Jaco Finlay were stored in a cardboard box up on the third floor.

"I told the people we have to do something; no one's Honorable Ancestor belongs in a box like that. But nothing happened. After a while I got in the car with my mother and daughter; I guess my little girl was nine years old that summer. We drove around to the reservations, looking for all the Finlays we could find, trying to drum up the interest to get him reinterred."

Everywhere they traveled they saw different versions of his name: Findlay Creek above Canal Flats in British Columbia; Findley Road on the Pend Oreille; the Jocko River in Montana.

"I remember coming to Finlay Point on Flathead Lake; we had ourselves a picnic there," Jeannette said. "It was nice to be at a place where we knew our ancestor had been. We talked to a lot of relatives, but it was like anything else—some people showed their concern, and some didn't. When we got back to Spokane I told the board that if they didn't do something, I was going to do it myself. After that things started to move."

Jeannette helped make the preparations for a small service on the grounds of the Spokane House State Park. The reinterment was scheduled for July 25, 1976, with a Jesuit priest and a Congregational minister presiding. Everyone was in accord, and several descendants planned to make the trip to the park for the simple ceremony.

The evening of July 24, a representative from the Park Service called to say that it would be illegal to bury anyone at Spokane House because the ground had not been designated as a cemetery. Someone on the museum board knew a judge, and by the time a modest crowd of museum staffers, Jesuit priests, and tribal members gathered the next afternoon, a document had been signed designating a little square of dirt off the edge of the old bastion as a legal burial site. They laid Jaco Finlay's remains in that place, exactly where they had dug them up.

"You know, the old people, they always said that Teshwentichina was the one who placed Jaco Finlay's knife and pipe bag in his grave," Jeannette said, "so everything we had went back." Jeannette recalled the moment clearly. "Bones and buttons, little pipe bowls, whatever else was in the box. We covered them up and said that's where it all belongs."

Acknowledgments

I INTERVIEWED, QUESTIONED, AND IMPOSED upon many people in the course of researching this book, and I would like to acknowledge some of the special assistance I received.

Dozens of archivists across the country responded to queries and retrieved obscure materials; although space prevents naming them all, their contributions are evident on every page. I am indebted to the Eastern Washington Historical Society and the staff of the Spokane Public Library, especially to Nancy Compeau in the Northwest Room and Mollie Coffee of the Inter-Library Loan Department.

Rolf Ludvigsen of the Denman Institute for Research on Trilobites introduced me to the Tanglefoot site and gave me excellent guidance; Ken Pugh, Gordon Moehse, and Dave Schaepe of the Sto:lo-Coast Salish Tribes helped sort out the story of Al Purvis's artifact collection.

Todd Thompson, Lisa Hallock, Bill Leonard, Tom Burke, Julie Fronzuto, and students of the Cusick School all shared their boundless enthusiasm for salamanders.

Rich Zack toured me through the entomological collection at Washington State University; Dennis Strenge and Patti Ensor told me how to look for sheepmoths in the wild; Jon Shepard and Don Rolfs helped unravel their taxonomy.

Glen Leitz, Charles T. Luttrell, and his parents Don and Edna generously shared their research on the Palouse mammoths; Jerry Galm and Stan Gough of Eastern Washington University showed us the correct site and helped piece together the bones of the story. Mary Hennen waded through hundreds of files at the Chicago Academy of Sciences, and Bill Simpson of the Field Museum of Natural History climbed a ladder to get just the right shot. Tina Wynecoop loaned me her mammoth molar from Foulweather Bluff.

David DeSante recounted details of a singular condor incident, and Lloyd Kiff of the World Center for Birds of Prey provided context for the recent history of the species. Virginia Butler of Portland State University enlightened me on the fishes of the Dalles Road Cut, and Victoria Hansel-Kuehn graciously laid out her bird bones for me.

Lil Fenn and Robert Boyd greatly broadened my understanding of the western spread of smallpox.

Karl Wegmann, of the Geology Division of the Washington Department of Natural Resources, patiently explained the complex interfingerings of the lower Columbia landscape.

Pam Camp told me where to go looking for native tobacco, then tromped through the sagebrush with me. Peter Lesica and Rich Old toured me through the herbariums of the Northwest. Patrick Left Hand and Wilfred Jacobs gently told me where to stop, and Jack Linnville gave me the run of his ranch.

Roy Breckenridge convinced me of the importance of Gable Mountain. Steve Reidel, geologist, and Darby Stapp, archaeologist, both of Pacific Northwest National Laboratory, placed the mountain in its natural and human context. Gladys Para brought its story up to modern times.

Alice Ignace took me up to her mountains; Dean Osterman of the Kalispel Tribe explained the role of language in culture; Allan Smith, no longer with us, gathered information that will be used for years to come.

Mark Weadick introduced me to the muskrat world; Steve and Gene Schalock showed me how they saw it; Roger Scheurer explained the business end. Pauline Flett, Ray Brinkman, and Father Tom Connolly, S.J., put the animals into a tribal context.

Chalk Courchaine opened up his extensive Finlay genealogy, the product of years of work. Jerry Peltier came up with a little black magic, and Albert Culverwell recalled the circumstances around the Finlay reinterment ceremony. Walt Goodman, Jeannette Whitford, Kay Hale, Jim Perkins, Lucille Otter, and many other descendants would make Jaco proud.

I would like to thank Ann Rittenberg and Gary Luke and the Sasquatch staff for seeing this book through. Ted McGregor of the *Pacific Northwest Islander* published early versions of several of these chapters.

Richard and Harriette Beckham, my brother Jim and my sister Mary, Holly and Walker Hillegass, Dennis Dauble, Charles Ferree, Joan Gregory, Dave Godlewski, Harriet Huber, Gene and Nancy Hunn, Bill Moreau, John Stern, and Jean Wood all listened to me try to explain these ideas, over and over again.

Claire served as my inspiration, microfilm reader, tireless sleuth, knuckle-rapper, and peerless red pencil. Emily and Jamie walked, waded, and waited, adding immeasurably to the fun of this project.

Selected Bibliography

The following primary sources were used in two or more chapters. Citations specific to each chapter follow.

Audubon, John James. *The Birds of America*. Vol. 4. New York: J. J. Audubon, 1840.

Cooper, James G., and George Suckley. *The Natural History of Washington Territory and Oregon*. New York: Bailliere Brothers, 1860.

Cox, Ross. *The Columbia River*. 1831. Reprint, edited by Edgar and Jane Stewart, Norman: University of Oklahoma Press, 1957.

Douglas, David. *Journal Kept by David Douglas During His Travels in North America, 1823–1827*. London: William Wesley & Son, 1914.

Douglas, David. "A Sketch of a Journey to the North-Western Parts of the Continent of North America, during the Years 1824, 5, 6, and 7." *Companion to the Botanical Magazine* 2 (1836): 79–182.

Henry, Alexander. *The Journals of Alexander Henry the Younger*. Vol. 2. Edited by Barry Gough. Toronto: The Champlain Society, 1992.

Hunn, Eugene S. *Nch'i-Wana: Mid-Columbia Indians and Their Land*. Seattle: University of Washington Press, 1990.

Landerholm, Carl, ed. *Notices and Voyages of the Famed Quebec Mission to the Pacific Northwest*. Portland: Oregon Historical Society, 1956.

Moulton, Gary E., ed. *The Journals of the Lewis and Clark Expedition*. Lincoln: University of Nebraska Press, 1983–1993.

Parker, Samuel. *Journal of an Exploring Tour Beyond the Rocky Mountains*. Moscow: University of Idaho Press, 1990.

Ross, Alexander. *Adventures of the First Settlers on the Oregon or Columbia River*. London: Smith, Elder, and Co., 1849.

Scouler, John. "Journal of a Voyage to North West America." *Oregon Historical Society Quarterly* 6 (1905): 159–205; 276–287.

Simpson, George. *Fur Trade and Empire: George Simpson's Journal . . . 1824–1825*. Edited by Frederick Merk. Cambridge: Harvard University Press, 1968.

Thompson, David. Journals. Provincial Archives of Ontario, Toronto.

Thompson, David. *Narrative of Explorations in Western North America, 1784– 1812*. Edited by J. B. Tyrrell. Toronto: Champlain Society, 1916.

Townsend, John Kirk. *Narrative of a Journey across the Rockies to the Columbia*. 1839. Reprint, with an introduction by Donald Jackson, Lincoln: University of Nebraska Press, 1978.

Vancouver, George. *A Voyage of Discovery to the North Pacific Ocean and Round the World, 1791–1795*. Edited by William Kaye Lamb. London: Haklyut Society, 1948.

Chapter 1: Little Stone House

Chatterton, Brian, and Rolf Ludvigsen. *Upper Steptoean (Upper Cambrian) Trilobites from the McKay Group of Southeastern British Columbia, Canada*. The Paleontological Society, Memoir 49. Lawrence, Kansas, 1998.

Fortey, Peter. *Trilobite! Eyewitness to Evolution*. New York: Alfred A. Knopf, 2000.

Ludvigsen, Rolf. *Life in Stone: A Natural History of British Columbia's Fossils*. Vancouver: University of British Columbia Press, 1996.

Ludvigsen, Rolf. "Myth, Magic, and Folklore of Fossils." *British Columbia Paleontological Alliance Newsletter* , n.d.

Oakley, Kenneth. "Folklore of Fossils, Part 2." *Antiquity* 39 (1965): 117–125.

Royal Tyrrell Museum of Paleontology. *The Land Before Us: The Making of Ancient Alberta*. Edmonton, Alberta: Lone Pine Publishing, 1994.

Taylor, Michael E., and Richard A. Robison. "Trilobites in Utah Folklore." *Brigham Young University Geology Studies* 23 (July 1976): 1-5.

Whitehouse, F. W. "The Australian Aborigine as a Collector of Fossils." *Queensland Naturalist* 13 (March 1948): 100–102.

Whittington, Henry B. *Fossils Illustrated: Trilobites*. Rochester, N.Y.: Boydell Press, 1992.

Chapter 2: Water Dogs

Black, Samuel. *Faithful to Their Tribe and Friends: Samuel Black's 1829 Fort Nez Perces Report*. Edited by Dennis W. Baird. Northwest Historical Manuscript Series. Moscow: University of Idaho Library, 2000.

Fronzuto, Julie Ann. "Predation and Antipredator Tactics in Ambystomatid Salamanders." Ph.D. diss., Washington State Univeristy, 2000.

Gao, Ke–Qin, and Nell H. Shubin. "Late Jurassic Salamanders from Northern China." *Nature* 410 (29 March 2001): 574–576.

Hines, Donald M. *Ghost Voices*. Issaquah, Wash.: Great Eagle Publishing, 1992.

Leonard, William P., et. al. *Amphibians of Washington and Oregon*. Seattle: Seattle Audubon Society, 1993.

Petranka, James W. *Salamanders of the United States and Canada*. Washington, D.C.: Smithsonian Institution Press, 1998.

Sahagun, Bernardino de. *Florentine Codex.* Translated by Charles E. Dibble and Arthur J. O. Anderson. Monographs of the School of American Research and the Museum of New Mexico, no. 14, part 12. Santa Fe, 1963.

Slater, James R. "*Ambystoma Tigrinum* in the State of Washington." *Copeia* 4 (31 December 1934): 198.

"The Medical Lake: What Is It." *Spokane Morning Review*, 15 July 1886.

Chapter 3: WHITE SHIELD

Aldrich, J. M. "Larvae of a Saturniid Moth Used as Food by California Indians." *Journal of the New York Entomological Society* 20 (March 1912): 28–31.

Corning, Howard, ed. *Letters of John James Audubon, 1826 – 1840*. Boston: Club of Odd Volumes, 1930.

Dana, Richard Henry. *Two Years Before the Mast*. New York: Dodd, Mead, 1946.

Ferguson, Douglas C. *The Moths of America North of Mexico*. London: E. W. Classey Limited and R. B. D. Publications, 1971.

Graustein, Jeannette E. *Thomas Nuttall, Naturalist*. Cambridge: Harvard University Press, 1967.

Harris, Thaddeus W. *A Report on the Insects of Massachusetts, Injurious to Vegetation*. Cambridge, Mass.: Folsom, Wells, & Thurston, 1841.

Lee, Daniel, and J. H. Frost. *Ten Years in Oregon*. New York: J. Collord, 1844.

Peale, Titian Ramsay. "Lepidoptera Americana." Peale Collection. Academy of Natural Sciences, Philadelphia.

Strecker, Herman. *Lepidoptera, Rhopaloceres, and Heteroceres, Indigenous and Exotic*.

Reading, Pa.: Owen's Steam Book and Job Printing Office, 1872–1877.

Strenge, Dennis. "Life-cycle of *Hemileuca hera hera* (Saturniidae) in the Columbia Basin." *Northwest Lepidopterist's Association Newsletter*, Spring 1997.

Strenge, Dennis and Richard Zack. "Observations on the Life History of the Sagebrush Sheepmoth." *Northwest Science* 75, no. 2 (2001): 118–121.

Tuskes, Paul M., James P. Tuttle, and Michael M. Collins. *The Wild Silk Moths of North America*. Ithaca: Cornell University Press, 1996.

Wyeth, Nathaniel J. *The Correspondence and Journals of Captain Nathaniel J. Wyeth.* Edited by Frederick G. Young. 1899. Reprint, New York: Arno Press, 1973.

Chapter 4: THE BEAUTIFUL BUZZARD OF THE COLUMBIA

Aoki, Haruo. *Nez Perce Dictionary*. Berkeley: University of California Press, 1994.

Audubon, John James. *Ornithological Biography.* Vol. 4. Edinburgh: Adam and Charles Black, 1838.

Baird, Spencer F. *The Birds of North America.* Philadelphia: J. B. Lippincott, 1860.

Barnston, George. "Abridged Sketch of the Life of Mr. David Douglas, Botanist, with a Few Details of his Travels and Discoveries." *Canadian Naturalist and Geologist* 5 (1860).

Boas, Franz. *Chinook Texts*. Washington, D.C.: Government Printing Office, 1894.

Cressman, L. S. "Cultural Sequences at The Dalles, Oregon." *Transactions of the American Philosophical Society* 50, part 10 (1960).

Dall, William H. *Spencer Fullerton Baird: A Biography*. Philadelphia: J. B. Lippincott, 1915.

Douglas, David. "Observations on the *Vultur Californianus* of Shaw." *Zoological Journal* 4 (1829): 328–330.

Fannin, John. "The California Vulture in Alberta." *The Auk* 14 (1897): 89.

Finley, William L. "Life History of the California Condor." *The Condor* 10 (January–February 1908): 5–10.

Fisher, H. I. "The Skeletons of Recent and Fossil Gymnogyps." *Pacific Science* 1 (1947): 227–236.

Grinnell, Joseph. "Archibald Menzies, First Collector of California Birds." *The Condor* 34 (November 1932): 243–252.

Hansel–Kuehn, Victoria. "The Dalles Roadcut (Fivemile Rapids) Avifauna: Evidence for a Cultural Origin." Master's thesis, Washington State University, 2003.

Hargreaves, Sheba. "The Letters of Roselle Putnam." *Oregon Historical Quarterly* 29 (1928): 242–264.

Harris, Harry. "The Annals of Gymnogyps to 1900." *The Condor* 43 (January– February 1941): 3–55.

Hines, Donald M. *Ghost Voices*. Issaquah, Wash.: Great Eagle Publishing, 1992.

Kiff, Lloyd. *The California Condor Recovery Programme*. Boise, Idaho: The Peregrine Fund, 2000.

Mattina, Anthony. *Colville–Okanagan Dictionary*. University of Montana Occasional Papers in Linguistics, no. 5. Missoula, 1987.

Miller, Loye. "Bird Remains from an Oregon Indian Midden." *The Condor* 59 (January 1957): 59–63.

Peale, Titian Ramsay. *Diary of Titian Ramsay Peale: Oregon to California*. Edited by Clifford Merrill Drury. Los Angeles: Glen Dawson, 1957.

Peale, Titian Ramsay. "Zoology." Peale Collection. Academy of Natural Sciences, Philadelphia.

Schaeffer, Claude. "Was the California Condor Known to the Blackfeet Indians?" *Journal of the Washington Academy of Sciences* 41 (June 1951): 181–191.

Schlick, Mary Dodds. *Columbia River Basketry*. Seattle: University of Washington Press, 1994.

Snyder, Noel, and Helen Snyder. *The California Condor: A Saga of Natural History and Conservation*. San Diego: Academic Press, 2000.

Townsend, John Kirk. "Popular Monograph of the Accipitrine Birds of N. A., No. 2." *The Literary Record and Journal of the Linnaean Association of Pennsylvania College* 4 (October 1848): 249–55; 265–72.

Townsend, John Kirk. *Ornithology of the United States of North America*. Philadelphia: J. B. Chevalier, 1839.

Turner, Harriet. *Ethnozoology of the Snoqualmie*. 2nd ed. N.p.: H. Turner, c. 1976.

Wagner, Henry R. *Spanish Voyages to the Northwest Coast of America in the Sixteenth Century*. San Francisco: California Historical Society, 1929.

Wetmore, Alexander. "Avifauna of the Pleistocene in Florida." *Smithsonian Miscellaneous Collections* 85 (1931): 25–31.

Wetmore, Alexander. "The California Condor in Texas." *The Condor* 35 (January, 1933): 37–38.

Wilcox, T. E. "Occurrence of California Vulture in Idaho." *Journal of the Washington Academy of Science* 8 (1918): 25.

Chapter 5: BEHEMOTH

Condon, Thomas. "The Fossil Mammoth of the Columbia Basin." *West Shore* 4 (1877): 72.

Donald, James T. "Notes on Elephant Remains from Washington Territory." *Canadian Naturalist* 9 (1879): 53–56.

Edwards, Jonathan. *An Illustrated History of Spokane County*. [San Francisco]: W. H. Lever, 1900.

Erickson, Edith E. *Colfax, 100 Plus*. [Colfax, Wash.: E. E. Erickson], c. 1981.

"Folklore of the Mammoth." *Popular Science Monthly* 21 (May 1882): 137.

Gilbert, Frank T. *Historic Sketches of Walla Walla, Whitman, Columbia, and Garfield Counties*. Portland: Printing House of A. G. Wlling, 1882.

Hay, Oliver. *The Pleistocene of the Western Region of North America and Its Vertebrate Animals*. Washington, D. C.: Carnegie Institution, 1927.

Haynes, Gary. *Mammoths, Mastodonts, and Elephants*. Cambridge University Press, 1991.

Higley, William K. "A Paper on *Elephas Primigenius*." *Bulletin of the Chicago Academy of Science* 1 (1886): 123–127.

Lister, Adrian, and Paul Bahn. *Mammoths*. London: Boxtree, 1995.

Luttrell, Charles T. "Three Fossil Discoveries in Washington Territory and Their Histories." Master's thesis, Eastern Washington University, 2001.

Matzger, William O. "Fossil Elephant in Washington Territory." *American Journal of Science and Arts* 13 (February 1877): 157.

Osborn, Henry F. "Paraelephas in Relation to Phyla and Genera of the Family Elephantidae." *American Museum Novitates* 152 (20 December 1924): 1–7.

Osborn, Henry F. *Proboscidea*. Vol. 2. New York: American Museum Press, 1936.

Osborn, Henry F. "New Subfamily, Generic, and Specific Stages in the Evolution of the Proboscidea." *American Museum Novitates* 99 (27 December 1923): 1–4.

Osborn, Henry F. "Species of American

Pleistocene Mammoths, *Elephas Jeffersonii*, New Species." *American Museum Novitates* 41 (8 July 1922): 1–16.

Shaver, F. A., Richard F. Steele, and A. P. Rose. *Illustrated History of Southeastern Washington.* Spokane: Western Historical Publishing Co., 1906.

Sternberg, Charles H. "*Elephas Columbi* and Other Mammals in the Swamps of Whitman County, Washington." *Science* 17 (27 March 1903): 511–512.

Sternberg, Charles H. "Experiences with Early Man in America." *Transactions of the Kansas Academy of Science* 18 (1903): 89–93.

Sternberg, Charles H. "Explorations in Northeastern Oregon." *Kansas City Review of Science* 7 (1884): 674–678.

Sternberg, Charles H. *The Life of a Fossil Hunter.* New York: Holt & Co., 1909.

Sternberg, Charles H. "The Quaternary of Washington Territory." *Kansas City Review of Science and Industry* 4 (1881): 601–602.

Chapter 6: THE DEVOURING DISORDER

Boyd, Robert T. *The Coming of the Spirit of Pestilence.* Vancouver: University of British Columbia Press, 1999.

Fenn, Elizabeth A. *Pox Americana.* New York: Farrar, Strauss and Giroux, 2001.

Fenner, F., et. al. *Smallpox and Its Eradication.* Geneva: World Health Organization, 1988.

Fidler, Peter. Journals. Hudson's Bay Company Archives, Provincial Archives of Manitoba, Winnipeg.

Harris, Cole. "Voices of Disaster: Smallpox around the Strait of Georgia in 1782." *Ethnohistory* 41 (Fall 1994): 591–626.

McDougall, Duncan. "Astoria Journal, 1810–13." Rosenbach Library, Philadelphia.

Mengarini, Gregory. *Recollections of the Flathead Mission.* Edited by Gloria Ricci Lothrop. Glendale, Ca.: Arthur H. Clark, 1977.

"Red Men Buried Many Years Ago." *Spokesman–Review,* 22 January 1905.

Rich, Edwin E., ed. *Cumberland House Journals and Inland Journal, 1775–1782.* London: Hudson's Bay Record Society, 1952.

Schaeffer, Claude E. Papers. Glenbow Museum Archives, Calgary, Alberta.

Schaeffer, Claude E. "Plains Kutenai: An Ethnological Evaluation." *Alberta Historical*

Review 30 (1982): 1–9.

Smith, Asa Bowen. Letter to Reverend D. Greene. Kamiah, Oregon Territory, 6 February 1840. In *The Diaries and Letters of Henry H. Spaulding and Asa Bowen Smith Relating to the Nez Perce Mission, 1838–1842,* edited by Clifford Drury, 136–37. Glendale, Ca.: Arthur H. Clark, 1958.

Work, John. "Fort Colvile Report." Hudson's Bay Company Archives, Provincial Archives of Manitoba, Winnipeg.

Chapter 7: MOUNT COFFIN

Allan, George. "Reminiscences of Fort Vancouver." In *Transactions of the Ninth Annual Reunion of the Oregon Pioneer Association for the Year 1881,* 75–80. Salem: E. M. Waite, 1882.

Barker, Burt Brown, ed. *Letters of Dr. John McLoughlin.* Portland: Binford & Mort, 1948.

Belcher, Edward. *Narrative of a Voyage Round the World.* London: Henry Colburn, 1843.

Boyd, Robert. *The Coming of the Spirit of Pestilence.* Vancouver: University of British Columbia Press, 1999.

Carter, Tolbert. "Pioneer Days." In *Transactions of the Thirty-Fourth Annual Session of the Oregon Pioneer Association,* 65–103. Portland: Peaslee Brothers, 1907.

Crawford, P. W. "Narrative of the Overland Journey to Oregon." Bancroft Collection. University of California, Berkeley.

Douglas, David. "Second Journey to Northwestern Parts of the Continent of North America." *Oregon Historical Quarterly* 6 (1905): 288–309; 417–449.

Gibbs, George. "Account of Indian Mythology in Oregon and Washington Territories." *Oregon Historical Quarterly* 56 (1955): 293–325.

Howay, Frederic W. "The Brig *Owyhee* on the Columbia, 1829–1830." *Oregon Historical Quarterly* 35 (1934): 10–21.

Howay, Frederic W., and T. C. Elliott. "Vancouver's Brig *Chatham* in the Columbia." *Oregon Historical Quarterly* 43 (1942): 318–327.

Kane, Paul. *Wanderings of an Artist among the Indians of North America.* 1859. Reprint, Edmonton: M. G. Hurtig, 1986.

Moore, Carlton. "Historic Coffin Rock." *Longview Daily News,* 17 June 1954.

Morton, Samuel George. *Crania Americana.*

Philadelphia: J. Dobson, 1839.

Ogden, Peter Skene. *Traits of American Indian Life.* 1853. Reprint, Fairfield, Wash.: Ye Galleon Press, 1986.

Olson, Mrs. Charles. *Cowlitz County, Washington, 1854–1948.* N.p., n.d.

Ott, Ruth, and Dorothy York. *History of Cowlitz County, Washington.* Kelso, Wash.: Cowlitz County Historical Society, c. 1983.

Rich, Edwin E., ed. *The Letters of John McLoughlin from Fort Vancouver to the Governor and Committee.* Toronto: The Champlain Society, 1941.

Scouler, John. "Remarks on the Form of the Skull of the North American Indian." *Zoological Journal* 4 (1829): 304–309.

Sturtevant, William, ed. *Handbook of North American Indians.* Vol. 7, *Northwest Coast,* edited by Wayne Suttles. Washington, D.C.: Smithsonian Institution Press, 1990.

Swan, James G. *The Northwest Coast, or, Three Years' Residence in Washington Territory.* 1857. Reprint, Seattle: University of Washington Press, 1972.

Taylor, Herbert C., and Lester L. Hoaglin. "Intermittent Fever Epidemic of the 1830s on the Lower Columbia River." *Ethnohistory* 9 (1962): 160–78.

Thornton, J. Quinn. *Oregon and California in 1848.* New York: Harper & Brothers, 1849.

Tolmie, William Fraser. *Physician and Fur Trader.* Vancouver, B. C.: Mitchell Press, 1963.

Victor, Frances F. "Flotsom and Jetsom of the Pacific." *Oregon Historical Quarterly* 2 (1901): 36–54.

Warre, Henry J. *Overland to Oregon in 1845.* Ottawa: Public Archives of Canada, 1976.

Wilkes, Charles. "Diary of Wilkes in the Northwest." *Washington Historical Quarterly* 16 (1925) and 17 (1926).

Wilkes, Charles. *Narrative of the United States Exploring Expedition During the Years 1838, 1839, 1841, 1842.* Vol. 5. Philadelphia: Lea and Blanchard, 1845.

Chapter 8: SMOKE

Chamberlain, Alexander B. "Ethnobotany of the Kootenay Indians." Chamberlain Papers. National Anthropological Archives, Smithsonian Institution, Washington, D.C.

Cusick, William. Notes. Washington State University Herbarium, Pullman.

Goodrich, Thomas Harper. *The Genus Nicotiana.* Waltham, Ma.: Chronica Botanica Co., 1954.

Gunther, Erna. *Ethnobotany of Western Washington.* Seattle: University of Washington Press, 1973.

Hill–Tout, Charles. "The Moses Coulee Pipe." *Transactions of the Royal Society of Canada* 29 (1935): 219–224.

Hodge, Frederick Webb, ed. *Handbook of American Indians North of Mexico.* Smithsonian Institution, Bureau of American Ethnology Bulletin 30. Washington, D.C., 1910.

Howay, Frederic W., and T. C. Elliott. "Vancouver's Brig *Chatham* in the Columbia." *Oregon Historical Quarterly* 43 (1942): 318–327.

Howay, Frederick W., ed. *Voyages of the "Columbia" to the Northwest Coast.* Portland: Oregon Historical Society Press, 1990.

Ingraham, Joseph. *Journal of the Brigantine "Hope" on a Voyage to the Northwest Coast.* Edited by Mark D. Kaplanoff. Barro, MA: Imprint Society, 1971.

Lyman, R. Lee. "Zooarchaeology of the Moses Coulee Cave Spoils Pile." *Northwest Anthropological Research Notes* 29 (spring 1995): 141–176.

May, Alan G. "The Moses Coulee Pipe." *American Antiquity* 8 (1942): 166–69.

Menzies, Archibald. *Journal of Vancouver's Voyage, April to October, 1792.* Edited by C. F. Newcombe. Archives of British Columbia Memoir 5. Victoria: W. H. Cullin, 1923.

Ross, Alexander. *The Fur Hunters of the Far West.* Edited by Kenneth A. Spaulding. Norman: University of Oklahoma Press, 1956.

Sapir, Edward. *Wishram Texts.* Publications of the American Ethnological Society, vol. 2. Leiden: Late E. J. Brill, 1909.

Setchell, William Albert. "Aboriginal Tobaccos." *American Anthropologist* 23 (October–December 1921): 397–415.

Spinden, Herbert J. *Tobacco Is American: The Story of Tobacco before the Coming of the White Man.* New York: New York Public Library, 1950.

Work, John. "Fort Colvile Report." Hudson's Bay Company Archives, Provincial Archives of Manitoba, Winnipeg.

Chapter 9: THE OTTER SWIMS

Anglin, Ron. *Forgotten Trails.* Edited by Glen W. Lindeman. Pullman: Washington State University Press, 1995.

Chatters, James C., Sarah K. Campbell, Grant D. Smith, and Phillip E. Minthorn, Jr. "Bison Procurement in the Far West: A 2,100-Year-Old Kill Site on the Columbia Plateau." *American Antiquity* 60 (October 1995): 751–763.

Daugherty, Richard D. "The Intermontane Western Tradition." *American Antiquity* 28 (October 1962): 144–150.

Fry, Willis E., and Eric Paul Gustafson. "Cervids from the Pliocene and Pleistocene of Central Washington." *Journal of Paleontology* 48 (March 1974): 375–386.

Gerber, M. S. "Legend and Legacy: Fifty Years of Defense Production at the Hanford Site." Richland, Wash.: Westinghouse Hanford Company, 1992.

Gustafson, Eric Paul. "The Vertebrate Faunas of the Pliocene Ringold Formation, South-Central Washington." University of Oregon Museum of Natural History Bulletin No. 23. Eugene, 1978.

Harris, Mary Powell. *Goodbye, White Bluffs.* Yakima, Wa.: Franklin Press, 1972.

Mehringer, Peter J. "Late-Quaternary Pollen Records from the Interior Pacific Northwest and Great Basin of the United States." In *Pollen Records of Late-Quaternary North American Sediments,* edited by V. M. Bryant, Jr. and R. G. Holloway. Dallas, Texas: American Associaton of Stratigraphic Palynologists, 1985.

Merriam, John C., and John P. Buwalda. "Age of Strata Referred to the Ellensburg Formation in the White Bluffs of the Columbia River." *University of California Publications, Bulletin of the Department of Geology* 10, no. 15 (1917): 255–266.

Orr, Elizabeth L., and William N. Orr. *Geology of the Pacific Northwest.* New York: McGraw–Hill, 1996.

Parker, Martha Berry. *Tales of Richland, White Bluffs, and Kennewick.* Fairfield, Wash.: Ye Galleon Press, 1979.

Reidel, S. P., and N. P. Campbell. "Structure of the Yakima Fold Belt, Central Washington." In *Geologic Guidebook for Washington and Adjacent Areas,* edited by N. L. Joseph and others. Washington Division of Geology and Earth Resources Information Circular 86: 277–288. Olympia, 1989.

Relander, Click. *Drummers and Dreamers.* Seattle: Pacific Northwest National Parks & Forests Association, 1986.

Rickard, William H., and Linda D. Poole.

"Terrestrial Wildlife of the Hanford Site: Past and Future." *Northwest Science* 63 (1989): 184–193.

Sanger, S. L. *Working on the Bomb: An Oral History of WWII Hanford.* Portland: Portland State University, Continuing Education Press, 1995.

Symons, Thomas W. *Report of an Examination of the Upper Columbia River.* Washington, D.C.: Government Printing Office, 1882.

Tedford, R. H., and E. P. Gustafson. "First North American Record of the Extinct Panda *Parailurus.*" *Nature* 265 (17 February 1977): 621–623.

U.S. Department of the Interior. Office of Indian Affairs. *Report on Source, Nature and Extent of the Fishing, Hunting and Miscellaneous Related Rights of Certain Indian Tribes in Washington and Oregon.* Los Angeles, July 1942.

Chapter 10: A GOOD DAY FOR DIGGING ROOTS

Carriker, Robert C. *The Kalispel People.* Phoenix: Indian Tribal Series, 1973.

Chittenden, Hiram M., and Alfred T. Richardson. *Life, Letters, and Travels of Father Pierre-Jean DeSmet, S. J., 1801–1873.* 4 vols. New York: Francis P. Harper, 1905.

Davis, Leslie B. *Remnant Forms of Folk Narrative Among the Upper Pend d'Oreille Indians.* Anthropology and Sociology Papers No. 31. Missoula: University of Montana, 1957.

Diomedi, Alexander. *Sketches of Indian Life in the Pacific Northwest.* 1894. Reprint, Fairfield, Wash.: Ye Galleon Press, 1998.

Fahey, John. *The Kalispel Indians.* Norman: University of Oklahoma Press, 1986.

Giorda, J. "Dictionary of the Kalispel or Flathead Indian Language." Oregon Province Jesuit Archives, Gonzaga University, Spokane.

Mengarini, Gregory. *Recollections of the Flathead Mission.* Edited by Gloria Ricci Lothrop. Glendale, Ca.: Arthur H. Clark, 1977.

Morgan, Lawrence R. "Kootenay-Salishan Linguistic Comparison: A Preliminary Study." Master's thesis, University of British Columbia, 1980.

Smith, Allan H. *Archaeological Survey of the Pend Oreille River Valley from Newport to*

Jared, Washington, August, 1957. Olympia: Washington State Department of Conservation, February 1958.

Vogt, Hans. *The Kalispel Language.* Oslo: I Kommisjon Hos Jacob Dybwad, 1940.

Chapter 11: MUSQUASH

Andrews, Antoine. Audiotape. Transcribed by Pauline Flett. Spokane Tribal Archives, Wellpinit, Washington.

Franchere, Gabriel. *Journal of a Voyage on the North West Coast of North America during the Years 1811, 1812, 1813, and 1814.* Edited and translated by W. Kaye Lamb and Wessie Lamb. Toronto: Champlain Society, 1969.

Gibbs, George. "Account of Indian Mythology in Oregon and Washington Territories." *Oregon Historical Quarterly* 56 (1955): 293–325.

Kennedy, Alexander. "Spokane House Report, 1822–23." Hudson's Bay Company Archives, Provincial Archives of Manitoba, Winnipeg.

Reichard, Gladys A. "Coeur d'Alene." In *Handbook of American Indian Languages.* Vol. 3. New York: J. J. Augustine, 1888.

"Spokane District Journal, From 15 April 1822 to 20 April 1813." Hudson's Bay Company Archives, Provincial Archives of Manitoba, Winnipeg.

Storer, Tracy I. "The Muskrat as Native and Alien." *Journal of Mammalogy* 18 (1937): 443–460.

Teit, James A. *Folk-Tales of Salishan and Sahaptin Tribes.* Lancaster, Pa.: American Folk-Lore Society, 1917.

Teit, James A. *The Salishan Tribes of the Western Plateaus.* 1927. Reprint, Seattle: Shorey Book Store, 1973.

Chapter 12: REBURYING JACO FINLAY

Caywood, Louis R. "Spokane House." The Beaver 281 (Winter 1956): 44–47.

Chittenden, Hiram M., and Alfred T. Richardson. *Life, Letters, and Travels of Father Pierre-Jean DeSmet, S. J., 1801–1873.* 4 vols.

New York: Francis P. Harper, 1905.

"Edmonton House Journals, 1828." Hudson's Bay Company Archives, Provincial Archives of Manitoba, Winnipeg.

Ferris, Joel E. "Early Outposts of the Fur Trade Empire." Ferris Collection. Eastern Washington Historical Society, Spokane.

"Finlay Family Tree." De Smetiana Collection. Jesuit Missouri Province Archives, St. Louis.

Harmon, Daniel. *Sixteen Years in the Indian Country.* Edited by W. Kaye Lamb. Toronto: Macmillan, 1957.

Jackson, John C. *Children of the Fur Trade: Forgotten Metis of the Northwest.* Missoula, Mont.: Mountain Press, 1995.

Jones, Robert F. *Annals of Astoria: The Headquarters Log of the Pacific Fur Company on the Columbia River, 1811 1813.* New York: Fordham University Press, 1999.

Masson, L. R., ed. *Les Bourgeois de la Compagnie du Nord-Quest.* 2 vols. 1889–90. Reprint, New York: Antiquarian Press, 1960.

McGillis Papers. Fur Trade Records. Archives of Ontario, Toronto.

Spokane House Files. Eastern Washington Historical Society, Spokane.

Thompson, David. "Narrative of the Establishment on the Scources of the Columbia." Royal Commonwealth Society Library, Cambridge.

Van Driel, John. "South Branch Massacre." Hudson's Bay Company Archives, Provincial Archives of Manitoba, Winnipeg.

Wallace, W. Stewart, ed. *Documents Relating to the North West Company.* Toronto: Champlain Society, 1934.

Work, John. "Journal of a Trip from Fort Colvile to Fort Vancouver and Return in 1828." *Washington Historical Quarterly* 11 (1920): 104–114.

Index

Index

Fraser River, 6, 160
Fur trading, 114, 189-90, 192, 194-98, 207-8, 210, 214-15

G

Gable, Hank, 161
Gable Mountain. *See also* Otter, The
    description of, 155, 158
    plutonium storage at, 164
    Trojan Nuclear Plant reactor vessel stored at, 166
Gass, Patrick, 44
Gibbs, George, 115
Ginkgo National Monument, 148
Glacier Park, 95
Goodman, Walt, 219-20, 222, 225
Gray, Robert, 112-13

H

Hanford, 162-63, 166
Hanford Reach National Monument, 166
Hangman Creek, 70, 73
Hansel-Kuehn, Victoria, 59
Harvey, Philip, 81
Hastings, Doc, 166
Herron, George, 223
Higley, W. K., 89
Horseshoe crab, 5
Howse Pass, 212
Hudson Bay Company, 135, 145
    muskrat trading, 195-96
    operating base for, 215-16
    smallpox epidemic, 97
Hut Creek, 112

I

Ice Age, 83
Ignace, Alice, 168, 170-72, 175-77
Indian(s). *See also specific tribe*
    diseases that affected, 107-8
    smallpox effects on, 95-98, 100, 104-5
Indian tobacco, 142-43
Iroquois, 210-11

J

Jaco Finlay. *See* Finlay, Jaco
Jocko River, 212
Johnson, J. M., 223

K

Kalispel Indians
    burial practices, 179-80
    epidemics, 179-80, 186
    historical descriptions of, 169-70
    Ignace's memories of, 168, 170-73, 181-82
    language of, 175, 178, 181, 183
    survival of, 180-81
    Thompson's contact with, 169
    Vogt's work with, 175-76, 180
Kalispel Reservation, 168-69
Kane, Paul, 124
Kenedy, J. H., 74
King, Mary Ann, 219-20, 223-25
King, Peter, 219, 224
Kingbirds, 36
Kootenae House, 209-10
Kootenai Indians
    Ignace's descriptions of, 173-74
    smallpox affliction of, 95, 106-8
    tobacco pipe smoking by, 135
Kootenay River, 5, 146, 212

L

La Grotte du Trilobite, 6-7
Labiesh, Francis, 43
LaDu, Crumline, 127-28
Lake Missoula, 111
Lake Pend Oreille, 212
Latah, 70, 72, 90
Lee, W. M., 89-90
Lewis, Meriwether
    California condor descriptions by, 42
    smallpox epidemics described by, 105
    tobacco smoking by, 136, 142
Little Spokane River, 213, 227
Louie, Joseph Grand, 224
Lucien, Etienne, 48-49
Ludvigsen Rolf, 6
Lyman, W. D., 88

M

Malaria, 121-23
Mammoth. *See* Columbian mammoth
McBride, J. J., 81
McDonald, Finan, 209, 213
McDonald, John, 207-8
McDougall, Duncan, 107
McKenzie, Donald, 140, 143
McLoughlin, John, 121
McMillan, James, 211
Medical Lake, 20-21
Mengarini, Gregorio, 175
Menzies, Archibald, 45-46, 134, 142
Merriam, C. Hart, 53, 62
Mole salamanders, 19
Morton, Samuel George, 120

larvae of, 16-17
long-toed, 13
mating of, 13, 15-16
metamorphosis of, 22-23, 25
misconceptions about, 11-12
mole, 19
neotenes, 25
raising of, 18-19
secrecy of, 13
tiger, 9, 15
Saleesh House, 212-13
Salish Arc, 175
Saskatchewan, 207-8
Saukamappee, 96-97
Saxifrage flowers, 2
Say's phoebes, 36
Schaeffer, Claude, 62, 64
Schalock, Steve, 189-90, 192
Scheurer, Roger, 193-94
Scouler, John, 117-18
Shannon, George, 43
Sheepmoth
  as caterpillars, 39-40
  Audubon's paintings of, 36-37
  author's experiences with, 30-32
  capturing of, 30-31
  description of, 28-29
  eggs of, 38
  fossil history, 37
  Indian depictions of, 38
  larvae of, 38-39
  mating of, 29-30
  metamorphosis of, 39-40
  mythological descriptions, 32
  Nuttall's collection of, 32-37
  sagebrush habitat of, 37-38
Shoshones, 140
Skillutes, 113
Slater, James, 21
Smallpox
  contagious nature of, 104
  death caused by, 103-4
  European resistance against, 104
  historical descriptions of, 94-95, 105-6
  Hudson Bay Company trading affected by,
    97, 99, 101-2
  Indian casualties caused by, 95-98, 100,
    104-5
  Kootenai Indians, 95-96, 106-8
  Lewis and Clark's writings about, 105-6
  New World affliction with, 104-5
  spread of, 106
  symptoms of, 103-4
  Tomison's descriptions of, 99-101

transmission of, 104
vaccinations against, 102-3
Walker's descriptions of, 97-99, 101-2
Smith, Asa, 94
Smoholla, 161-62
Snake River, 34, 38, 156, 215
Snake River Plain, 66
Snoqualmie tribe, 61
Spokane House
  artifacts of, 205-6
  Douglas' work at, 217
  excavation of, 204-5
  Jaco Finlay's stay at, 213, 215, 229
Spokane River, 204, 213, 217, 227
Steamboat Landing, 128
Sternberg, Charles H., 86
Stone House, 6-7
Suckley, George, 20, 52, 53
Swan, James, 120

T

Tanglefoot Creek
  description of, 2-3, 5
  trilobites in, 5-7
Teanaway River, 19
Tenochtitlan, 25
Teshwentichina, 213, 216, 229
Thompson, David
  description of, 97, 107
  Jaco Finlay's involvement with, 209-13
  Kalispel Indians contact with, 169
  muskrat pelts described by, 195
  The Otter descriptions, 159-60
  Rocky Mountain crossing by, 209
  tobacco descriptions by, 138-39
Thompson, Todd, 15
Thwing, Nathaniel, 84
Tobacco
  Broughton's descriptions of, 135, 144
  Columbia area cultivation of, 142, 145-46
  coyote, 142-43, 146, 148, 151
  cultivating of, 134
  Douglas' descriptions of, 140-42
  early uses of, 134-35
  economic benefits of, 144-45
  Fidler's descriptions of, 135-36
  historical descriptions of, 133-38
  Indian, 142-43
  Kootenai cultivation of, 145, 147
  Lewis and Clark's descriptions of, 136-38
  native peoples' use of, 134, 144-45
  plant descriptions, 132-34
  Shoshone use of, 140
  smoking of, 135-37

Jack Nisbet

JACK NISBET was raised in North and South Carolina, where his mother helped him pin insects and his grandmother introduced him to bird songs. He moved to eastern Washington after graduating from Stanford University in 1971, and for ten years he wrote a weekly natural history column for the *Chewelah Independent*.

His book *Sources of the River: Tracking David Thompson Across Western North America* (1994) was awarded the Murray Morgan History Prize. *Purple Flat Top* (1997) is a collection of short stories, and *Singing Grass, Burning Sage*, published in association with the Nature Conservancy of Washington, won a silver medal at the ForeWords 2000 book awards.

For the past ten years, Nisbet has lived with his wife and two children in Spokane while teaching and studying human and natural history around the region.